NUFFIELD
NATIONAL CURRICULUM
MATHEMATICS

Number and algebra 3

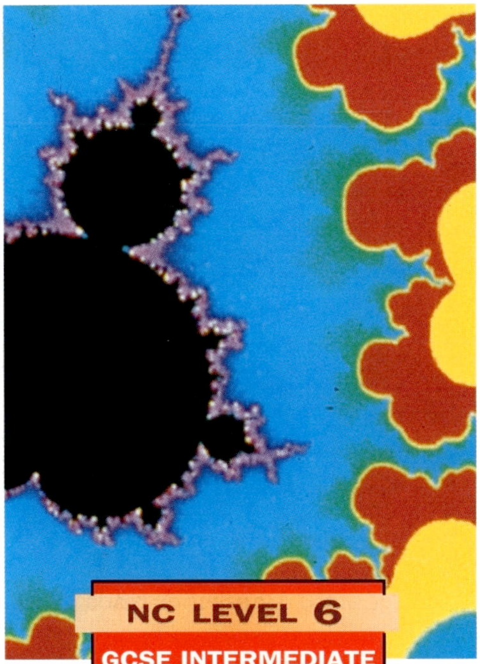

NC LEVEL 6
GCSE INTERMEDIATE

AF854608

Heinemann

Project director: Peter Reynolds
Assistant director: Mike Cornelius
Project manager: Jan Hoare

Authors of this book
Mike Cornelius and Peter Pool with additional material by Pat Perks and Stephanie Prestage
Edited by Mike Cornelius

Heinemann Educational
a division of Heinemann Publishers (Oxford) Ltd,
Halley Court, Jordan Hill, Oxford, OX2 8EJ

OXFORD LONDON EDINBURGH
MADRID ATHENS BOLOGNA PARIS
MELBOURNE SYDNEY AUCKLAND SINGAPORE TOKYO
IBADAN NAIROBI HARARE GABORONE
PORTSMOUTH NH (USA)

First published 1995

95 96 97 98 10 9 8 7 6 5 4 3 2

ISBN 0 435 50548 3

Designed and typeset by Ken Vail Graphic Design
Illustrated by Rupert Besley, Paul Cookson, Joan Corlass Nick Hawken, Brian Hoskin, Mike Lacey, Nigel Paige, Peter Wilks, Graeme Morris (Ken Vail Graphic Design)
Printed in Spain by Mateu Cromo, S.A. Pinto (Madrid)

Acknowledgements
The project and publishers gratefully acknowledge the contribution made by the late Hilary Shuard.
The authors and publishers would like to thank for permission to reproduce the following photographs and copyright material:
p. 6 Science Photo Library/Hank Morgan; p. 65 Collections/Brian Shuel;
p. 67 Colorsport.
Cover image: Science Photo Library/Gregory Sams
HMSO for permission to reproduce the National Curriculum level descriptions.

Cover picture
This fractal image is a visual way of showing an algebraic relationship from a group of complex numbers called the Mandelbrot Set. Unit 10 introduces ways of showing simple algebraic relationships visually in mappings and graphs.

About this book

This book is part of the Nuffield National Curriculum Mathematics series. It will help you learn mathematics and we hope you will enjoy using it.

How this book is organized

There are 12 units of work in this book. They are numbered Unit 1 to Unit 12. Each unit provides information for you to read and questions and activities for you to do.

The units are presented in short sections. Each section has a letter and a title. For example, Section B of Unit 2 has the title:

B Some investigations

The questions and activities in Section B are numbered **B1**, **B2**, **B3** and so on to help you find them easily

The contents list on the next page shows you where each unit and section starts and tells you what mathematics they cover.

Symbols used in the Nuffield books

These symbols are shown in the margin of your book whenever you need worksheets, equipment or access to a computer:

N3:4

this means you need Worksheet N3:4 from the Assessment and Resource Pack.

protractor
compass

you need equipment listed between the lines

spreadsheet

you need access to a computer and the software shown.

Contents

UNIT 1 *Accuracy with numbers*

A Trial and improvement

A square challenge

This square has

side = 3 cm

area = $(3 \times 3)\,\text{cm}^2 = 9\,\text{cm}^2$

This square has

side = 4 cm

area = $(4 \times 4)\,\text{cm}^2 = 16\,\text{cm}^2$

This square has an area of $11\,\text{cm}^2$.

Can you find w, the length of the side of the blue square?

The square challenge was given to Eva's class. They knew:

$3 \times 3 = 9$ $\quad 4 \times 4 = 16$ $\quad w \times w = 11$

How could they find w?

Eva said: $3 \times 3 = 9$ and $4 \times 4 = 16$ so w must lie between 3 and 4.

She tried $w = 3.5$. But $3.5 \times 3.5 = 12.25$ which is too big!

So how about 3.4×3.4?

A1 Do you think Eva made a good choice with $w = 3.4$? Give a reason for your answer.

George thought w might be about 3.2 so he tried:

$3.2 \times 3.2 = 10.24$ This is too small!

So how about 3.3×3.3?

A2 Who was closer to the right answer after two guesses, Eva or George?

Both pupils were determined to find a value for w which gave an area of 11 *exactly*. The only buttons they used on their calculators were the digits 0 and 9 and the [=] and [×] keys.

A3 See if you can find an exact value for w using just these keys.

When Eva and George compared results they found that Eva had made 21 guesstimates while George had made 18.

A4 **a** How many guesstimates did you make?

b Do you think you would obtain the same answer on any calculator? Try exchanging calculators to test your ideas.

c Do you think you have the exact answer now? Explain.

A5 Find the sides of squares which have areas (in square cm) of:

a 13 **b** 41 **c** 150

and in each case record the number of guesstimates you make.

A6 Two other pupils, Andrew and Denzil, thought they should be able to make use of other calculator buttons. They experimented using just the ten digit keys 0 to 9 and the keys:

[×] [x^2] [x^y] [=]

a Experiment with these keys to see what you can do with them.

b Now try to use these keys to obtain the sides of squares whose areas are:

i $2\,\text{m}^2$ **ii** $500\,\text{cm}^2$ **iii** $0.5\,\text{m}^2$ **iv** $0.1\,\text{m}^2$.

A7 Make an exact copy of this square:

a What is the area of the square?

b Using a different colour each time, colour in squares with areas:

i 9 cm^2 **ii** 11 cm^2 **iii** 15 cm^2

as accurately as you can and say what the lengths of the sides are in each case.

c Make another copy of your square. Repeat part **b** for squares with areas:

i 0.001 m^2 **ii** 0.002 m^2 **iii** 0.003 m^2

A8

How many $\frac{1}{2}$ metre squares will be needed?

A cube challenge

This cube has

side = 2 cm

volume = (2 × 2 × 2) = 8 cm^3

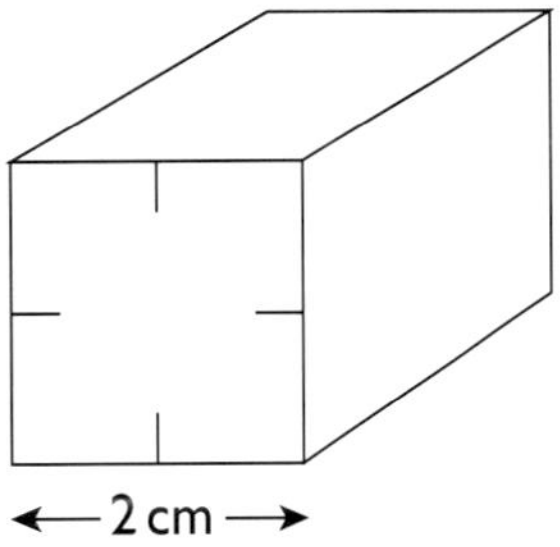

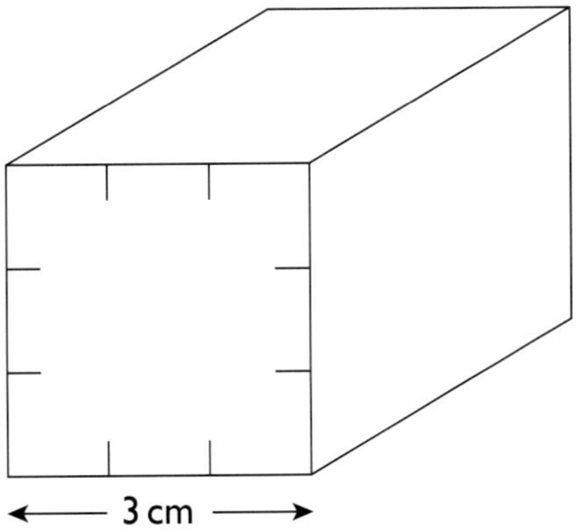

This cube has

side = 3 cm

volume = (3 × 3 × 3) = 27 cm^3

This cube has a volume of 11 cm^3.

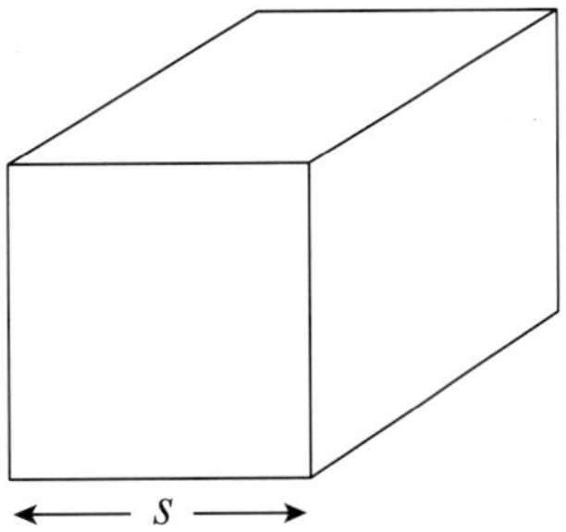

A9 Can you find the length s of the edge of the cube? Follow these stages:

a Between what two whole numbers must s be placed?

b See if you can find s using only the digits 0 to 9 and the [×] and [=] keys on your calculator.

c See if you can make use of the [x^y] key. Explain what you do.

d Can you make use of the [x^y] key in working out s?

A10 Work out the sides of cubes which have volumes (in cubic cm):

a 20 **b** 25 **c** 50 **d** 800 **e** 0.5

A11 What do you notice about your answer to **A10 e**? Try to explain.

A12 **a** You should already be able to use the key [$\sqrt{x}$] on your calculator but check by trying:

[9] [$\sqrt{x}$] or [$\sqrt{x}$] [9] (different calculators may use different orders).

b Use the [$\sqrt{x}$] key with the following numbers and record your answers:

i 16 **ii** 25 **iii** 100 **iv** 144
v 11 **vi** 500 **vii** 0.5 **viii** 0.1

c What does [$\sqrt{x}$] find if you have a square with an area of x square units?

> Finding the length of the side of a cube when you know the *volume* is called finding the **cube root**.
>
> The cube root symbol is $\sqrt[3]{}$.

A13 Some calculators have a short-cut key for a cube root: [$\sqrt[3]{x}$]
If your calculator has this key, use it to find the lengths of sides of cubes with the following volumes:

a $11\,cm^3$ **b** $20\,cm^3$ **c** $25\,cm^3$

d $50\,cm^3$ **e** $800\,cm^3$ **f** $0.5\,cm^3$

If your calculator does not have this key, find another way to work out the answers.

A14 **a** What is $2 \times 2 \times 2$?

b What is the cube root of 8?

c Try these calculator sequences and explain your results:

i [2] [x^y] [3] [=]

ii [8] [x^y] [(] [1] [÷] [3] [)] [=]

A15 **a** Find the cube roots of the numbers in question **A12** using the method just described.

b Try to explain why brackets are needed to obtain the correct answer.

c What happens if you do not include the brackets? Experiment and explain.

A16 Use a 'trial and improvement' method to find x if:

a $2x^2 = 13$ **b** $3x^3 = 13$ **c** $5x^2 = 23$ **d** $5x^3 = 23$

B Decimal places

You probably already know how to correct amounts to two decimal places. But sometimes we need greater accuracy.

The **micrometer** is an instrument used by toolmakers to measure lengths accurate to $\frac{1}{1000}$ or $\frac{1}{10\,000}$ of an inch.

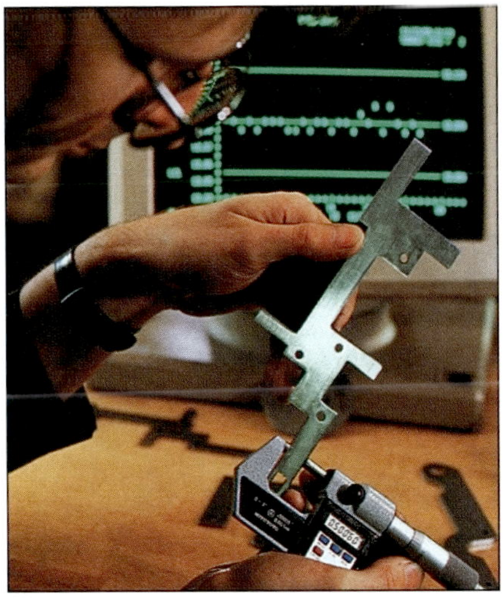

For example, a craftsman calculates that a piece of metal needs to be 3.67898 inches long. He has to be accurate to the nearest $\frac{1}{1000}$th of an inch (or three decimal places). So he corrects like this:

Units	$\frac{1}{10}$	$\frac{1}{100}$	$\frac{1}{1000}$	$\frac{1}{10\,000}$	$\frac{1}{100\,000}$
3	6	7	8	(9)	8

Since 9 ⩾ 5 it is large enough to change $\frac{8}{1000}$ to $\frac{9}{1000}$.

The craftsman now measures 3.679 inches with the micrometer – he has corrected the length to three decimal places.

B1 This diagram shows the floor plan of a classroom.

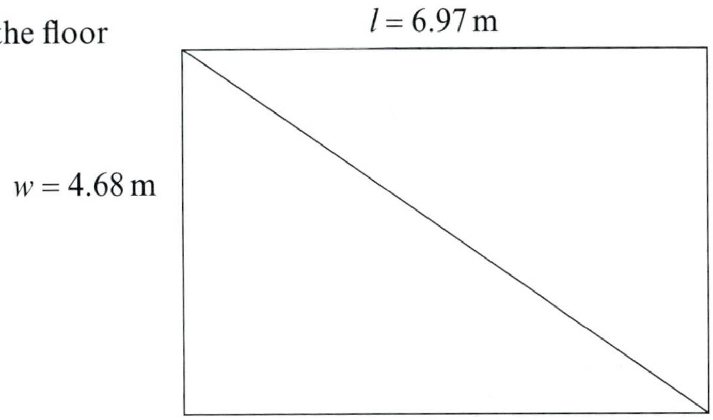

a If you draw the plan using a scale of 1 cm: 1 m, what will the length and width of the plan be?

b Correct the length and width to one decimal place and draw the plan as accurately as possible, using the scale in **a**.

c Measure the length of the diagonal and give your answer correct to one decimal place.

d What is the ratio of the measurements on the plan to the measurements of the actual classroom?

B2 Copy this table and fill in the last column:

Number	Corrected to one decimal place	True or false?
6.42	6.4	
3.78	3.7	
4.92	4.9	
4.08	4.1	
16.003	16.0	
5.009	5.1	
6.99	7.0	
5.95	6.0	
400.089	400.1	

B3 Correct these numbers to one decimal place:

a 8.53 **b** 4.19 **c** 3.94 **d** 4.85

e 17.918 **f** 19.455 **g** 12.98 **h** 13.0909

i 238.96 **j** 399.99

Some scientific calculators will correct numbers to a given number of decimal places. Find out if your calculator can do this.

B4 Copy and complete this table:

Number	No. of decimal places corrected to	Answer
3.78956	1	
3.78956	2	
3.78956	3	
3.78956	4	
4.07945	1	
4.07945	2	
4.07945	3	
4.07945	4	
329.99995	1	
329.99995	2	
329.99995	3	
329.99995	4	

What do you notice? Try to give a reason for your answers.

C Significant figures

Not all the digits in a number are of equal importance.

Consider £3.27

The 3 is the most **significant figure**.

C1 **a** If the amount was £3.99, would 3 still be the most significant figure? Give reasons for your answer.

b What other single number could represent £3.99 more accurately?

C2 Bronwen is owed £3.99 by her mother who has only £1 coins. How many coins should Bronwen get?

C3 It is 193 miles from Chichester to Lowestoft.

a Decide whether the following statements are true or false:

i correct to 1 significant figure the distance is 2 miles
ii correct to 1 significant figure the distance is 20 miles
iii correct to 1 significant figure the distance is 200 miles
iv correct to 2 significant figures the distance is 19 miles
v correct to 2 significant figures the distance is 190 miles.

b Discuss your results in **a** with a neighbour and see if you can work out a rule for correcting numbers to a required number of significant figures.

c Find Chichester and Lowestoft on a map. Give instructions to the pilot of an aeroplane on how to fly:

i from Chichester to Lowestoft **ii** from Lowestoft to Chichester.

C4 It is 309 km from Chichester to Lowestoft.

a Correct this distance to **i** 1 s.f. **ii** 2 s.f.
(Note: we often use s.f. as an abbreviation for significant figures.)

b Which answer in **a** is the most useful? Why?

C5 The actual cost of this boat is £59 899.50.

a At what price would you advertise it for sale?

b To how many significant figures have you corrected the price?

C6 Watch out for zeros!

Copy and complete this table:

When is a zero significant? Discuss this with a neighbour.

Number	Correct to 3 s.f.	True or false?
45692	45700	
4.5692	4.57	
456.92	457	
30208	302	
3.0208	3.02	
59999999	6000000	
59.999999	60.0	
3.20999	3.21	

C7 Consider 3.689759. Correct this number to:

a 2 d.p. **b** 2 s.f. **c** 3 d.p.

d 1 s.f. **e** 5 s.f. **f** 5 d.p.

D Negative numbers

D1 Copy and complete these tables:

2nd number (columns); 1st number (rows)

×	−2	−1	0	+1	+2
−2	+4		0		
−1				−1	
0					
+1					
+2					

2nd number (columns); 1st number (rows)

÷	−2	−1	0	+1	+2
−2	+1				
−1					
0				0	
+1		−1			+0.5
+2					

D2 From your results in **D1**, what can you say about:

a a minus times a plus

b a minus times a minus

c a minus divided by a plus

d a minus divided by a minus

e dividing by zero? (Try it with different numbers on your calculator.)

D3 In an examination a pupil gets 2 marks for a correct answer and –1 for a wrong answer. What marks will the following get:

a Paul: 13 right, 7 wrong **b** Pam: 9 right, 11 wrong

c Pat: 3 right, 17 wrong **d** Philip: 1 right, 19 wrong?

D4 If $a = -1$, $b = -2$, $c = 3$, $d = 4$ find:

a ab **b** $c \div b$ **c** $cd \times b$ **d** $3a \div c$

e $b \div a$ **f** $cd \div ab$ **g** $ac \div bd$

D5 A clock gains –3 minutes per hour.

a Rewrite this statement using a positive number.

b If the clock is correct at 12 noon, what time did it show at:
i 8 am **ii** 10 am **iii** 5 am?

c If the clock is correct at 12 noon, when will it be correct again? When was it last correct?

D6

What answers do they get?

D7

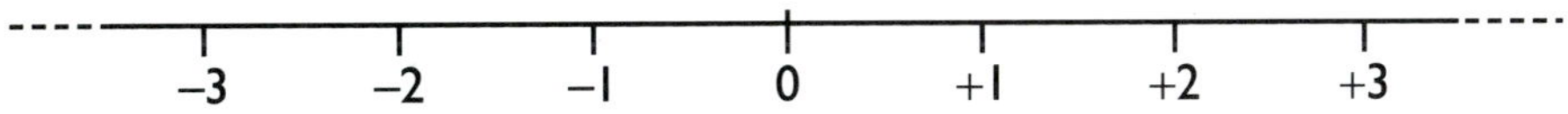

In a game with a number line, you start at 0. You throw a coin and a dice. Heads means '+' and Tails means '–'. So, for example, a throw of 'Tail, 5' would take you from 0 to –5 and you would move from –5 after the next throw.

a Where will you finish after throws of:

i H3, T2, T6, H1, T4, T2 **ii** T6, T5, H1, T3, H4, T4?

b Find sequences of throws which will leave you at:

i –10 after 6 throws **ii** +20 after 8 throws
iii –50 after 10 throws.

UNIT 2 Using numbers

A Fractions, decimals, ratios and percentages

Remember that fractions can be written in several different ways.
For example: $\frac{2}{5}$ means $2 \div 5$,
$2 \div 5 = 0.4$,
$0.40 = \frac{40}{100}$
and $0.4 = \frac{4}{10} = \frac{40}{100} = 40\%$.
These are all different versions of the same fraction.

A1 **a** Make a copy of this diagram with lines 12 cm long:

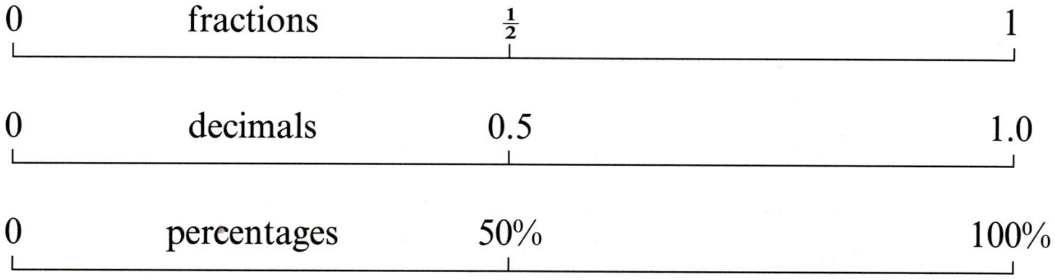

b On the first line of your diagram, write in the fractions $\frac{1}{4}$, $\frac{3}{4}$, $\frac{1}{3}$, $\frac{3}{5}$ and $\frac{5}{8}$ in their correct places.

c Use the fractions on the first line to estimate the corresponding decimals and percentages on the other two lines.

A **ratio** is often used instead of a fraction. For example:

To mix car radiator solution, use one part of antifreeze to three parts of water.

The ratio antifreeze : water may be written in different ways:

as **whole numbers**	1 : 3
in **fractions**	$\frac{1}{4} : \frac{3}{4}$
in **decimals**	0.25 : 0.75
in **percentages**	25% : 75%.

There is only *one way* to write amounts in decimals and percentages, but with ratios and fractions we could use **equivalent** ratios and **equivalent** fractions.

For example, 1:3 is the same as 2:6 or 3:9 or 25:75 and so on.

Similarly, $\frac{1}{4} = \frac{2}{8} = \frac{12}{48}$.

Usually we write these in their simplest forms. That is 1:3 and $\frac{1}{4}$.

To simplify a ratio such as 25:30, divide both numbers by 5 to give 5:6.

A2 Write these fractions

a $\frac{1}{4}$ **b** $\frac{2}{5}$ **c** $\frac{7}{8}$ **d** $\frac{45}{360}$ **e** $\frac{8}{9}$

as **i** decimals **ii** percentages.

A3 **a** How many days in September 1994 were Sundays?
Write this number as a percentage of the total days in the month.
Now write it as a fraction and as a decimal.

b Write down the ratio of number of Sundays : number of other days.

c What is the highest percentage of Sundays possible in *any* single month? When did this occur last? When will it occur next?

A4 Write these percentages

a 30% **b** $12\frac{1}{2}$% **c** 70% **d** 23% **e** 120%

as **i** fractions **ii** decimals.

A5 Write these ratios in their simplest forms:

a 2:10 **b** 5:100 **c** 23:460 **d** $\frac{1}{2}$:27

A6 A concrete mixer needs 4 parts sand to 1 part cement.

a What fraction of the mixture is sand?

b What is the fraction of sand as a decimal?

c What percentage of the mixture is cement?

d What is the ratio of cement to sand?

A7 At a vote, 30% of the voters voted 'Yes' and 70% voted 'No'.

a What was the ratio of 'Yes' votes to 'No' votes?

b What fraction voted 'Yes'? Write this as a decimal.

c If 600 voted 'Yes', this means that 30% of the total vote is 600.
We can now work out that 10% must be 200 and 70% (the 'No' votes) is $7 \times 200 = 1400$.
If 51 300 people voted 'Yes', how many voted 'No'?

A8 A painter mixes red, yellow and blue paint. The mixture contains red, yellow and blue in the proportion 1:2:5.

a What percentage of the mixture is yellow?

b What fraction of the mixture is red?

c If the mixture uses 3.7 litres of blue, how much red will it contain?

A9 In a school with 1200 pupils, 150 are left-handed.

a What is the ratio of left-handed to right-handed pupils in the school?

b What percentage of pupils are right-handed?

c What fraction of pupils are left-handed?

B Some investigations

Extension work

B1 How could you estimate how many pupils in your school are left-handed?

a First estimate what you think the percentage will be.

b Then collect information from a sample of about 50 pupils.

c On the basis of this sample, suggest what you think the total number of left-handed pupils might be.

B2 **a** A scientist believes that there are more left-handed women than men. Collect evidence to enable you to comment on this claim.

b Are cats and dogs more 'right-handed' than 'left-handed'? Devise a way of collecting evidence to help you to decide.

B3 Football teams seem to score more goals when they play at home than when they play away.

a Collect data from recent football matches and calculate the ratio of home goals : away goals.

b Express your results in terms of:
i fractions
ii percentages.

c Try the same investigation into your school's hockey, netball (or any other team) results from recent years. Do you get a similar ratio? If not, why do you think you get a different one?

B4 **a** Using a newspaper of your choice, work out the area covered by advertisements on a single page. Calculate the area of the whole page. You may find it more convenient to round your answers to suitable whole numbers.

b Calculate the percentage of space devoted to advertisements.

c Calculate the fraction of space devoted to advertisements.

d Calculate the ratio of advertisement space to other space.

e Compare your results with other people's newspapers. Represent the information for several newspapers on a diagram. Does this help you to decide a 'best buy'?

Proportion

B5 Watch a commercial television channel for exactly one hour.

a Use a table like the one below to keep a record of:

i time spent on advertisements

ii time spent on actual programmes

iii time spent on other things (e.g. previews of future programmes).

	Start	**Finish**	**Time taken**
Programme			
Advert 1			
Advert 2			

b Calculate the times in **i**, **ii** and **iii** as percentages and fractions of one hour.

c When three quantities are compared, the idea of ratio is extended to **proportion**. Write down the proportion **i:ii:iii**, using the smallest possible whole numbers.
You may need to round your answers to obtain a sensible answer.

d If the channel were to broadcast for 24 hours a day, calculate how many hours would be spent on **i**, **ii** and **iii** in a year if the proportions stayed the same. What assumptions do you have to make?

C Ratios and units

For every £1 spent, a supermarket will give 1p to charity.

This ratio could be written as:

1p : £1

but we normally try to have the same units on both sides of the ratio.

So we write: 1p : £1 = 1p : 100p

or: 1 : 100

without giving any units.

C1 A map has a scale 1 cm:1 km. Write this as a ratio with the same units.

C2 Simplify these ratios by converting to a common unit in each case:

a 1p : £5 **b** 1 cm : 1 m **c** 1 g : 3 kg **d** 1 second : 1 hour

e 3 mm : 8 cm **f** 1 inch : 1 mile **g** 7 g : 5 kg

C3 **a** Find three maps with different scales. For each of them, write down the scale in two different ways (such as 1 cm:1 km and 1:100 000). Take care: do *not* write 1 cm = 1 km.

b Suggest suitable scales for plans of:

i your house
ii your school
iii your nearest town
iv your county.

In each case, write down the scale in two different ways.

C4 An old map of the British Isles has a scale of 1:4 000 000.

a Express this scale in inches and miles.

b What area would 1 square inch on the map represent in square inches of actual ground?

D More uses for ratios

Here are some recipes:

Omelette (serves two people)

4 eggs
10 g butter
1 dessertspoon olive oil
2 medium potatoes
3 rashers bacon
1 large onion
50 g cheese

Casserole (serves six people)

275 ml stock
275 ml cider
25 g lard
8 carrots
10 button onions
12 slices bacon
100 g mushrooms
25 g plain flour

Oatmeal Porridge (serves two)

570 ml water
60 g medium oatmeal
1 teaspoon salt

Sponge Cake (two 18 cm tins)

110 g butter
110 g sugar
2 eggs
110 g flour

D1 Work out the quantities needed for the following:

a An omelette for five people. (Divide the amount given for each ingredient by 2 to find the amount needed for one person, then multiply by 5. Some answers may need to be adjusted.)

b A casserole for five people. (Find the amounts needed for one person and multiply by 5. You may need to adjust your answers.)

c Oatmeal porridge for three people.

d Sponge cake for seven tins.

Note down any problems you have and any assumptions you have to make.

Extension work

D2 An employer's weekly costs total £150 000.

a If these costs are made up of wages, materials and transport in the proportion 7:3:2, work out the cost of wages.

b What percentage of the total do transport costs represent?

c If wages increase by 3.5%, what is the new wage bill?

D3 Find out what an '18 cm tin' means. Then work out the quantities needed to make two 20 cm tins of Sponge Cake.

D4 Find a cookery recipe which caters for a given number of people. Write down and then calculate the quantities needed for:

a twice the number of people given

b two more than the number of people given

c one less than the number of people given.

Note down any problems you have in changing the recipe.

D5 Divide 24 into three parts in the proportion 3:4:5. The result is three numbers – a, b and c. Draw a triangle with sides a cm, b cm and c cm. What do you notice about the triangle?

D6 In a will, £1 million is left to three children, Ann, Bob and Clare, so that they receive $\frac{1}{7}$, $\frac{2}{7}$ and $\frac{4}{7}$ respectively. How much does each child receive? Express each share as a percentage.

D7 A Bedouin left his herd of 23 camels to his three sons so that one got one eighth, another one third and the last a half. Why did the sons have a problem sharing the camels?

A friend lent a camel (so there were now 24) and the sons divided the camels so that they took one eighth, one third and a half before giving back the camel to a friend. Can you explain what has happened? (Hint: fractions of what?)

E Enlarging

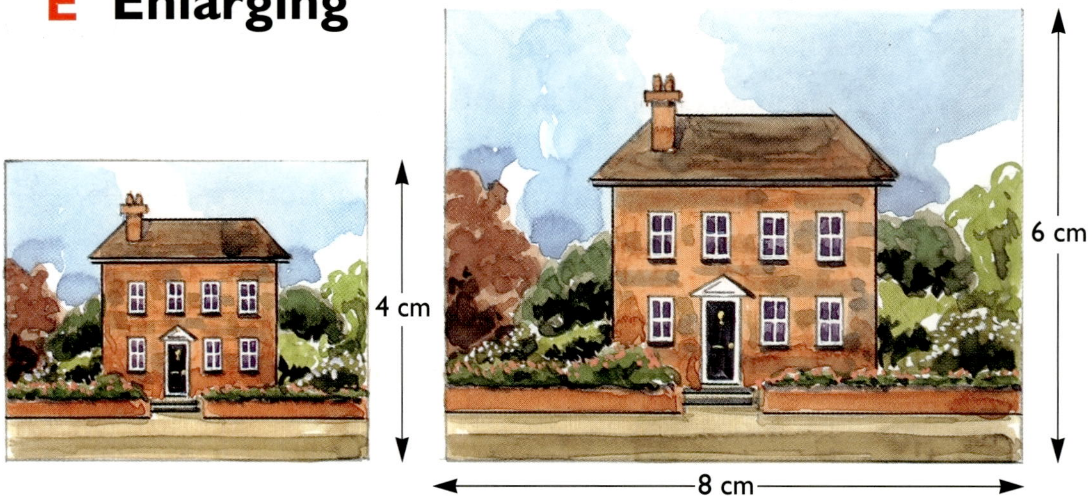

One of these pictures is an enlargement of the other.

Both pictures have the same **shape**. This means that the ratio of **corresponding lengths** in the pictures will be the same. We use **scale factor** to describe the change from one set of measurements to the other.

E1 The scale factor for the pictures is 1.5 because the height of 4 cm becomes 6 cm when it is multiplied by 1.5: $4 \times 1.5 = 6$.

a What is the ratio of the heights of the houses in the two pictures?

b What is the width of the small picture?

E2 A scale model of a car is 5 cm high. The actual car is 2 m high.

a What is the scale factor of the enlargement?

b What is the length of the model if the car is 4.4 m long?

c If the bonnet of the model is 3 cm long, how long is the real bonnet?

d What is the scale factor of enlargement for the area of the model to the area of the real car? Find the area of the roof of the car if the model has a roof of area 20 cm^2?

E3 If you made a scale model of yourself in plasticine, with a scale factor of $\frac{1}{5}$, what would be **a** the height, **b** the arm length and **c** the head circumference?

E4 The diagram shows the square ABDC being enlarged.

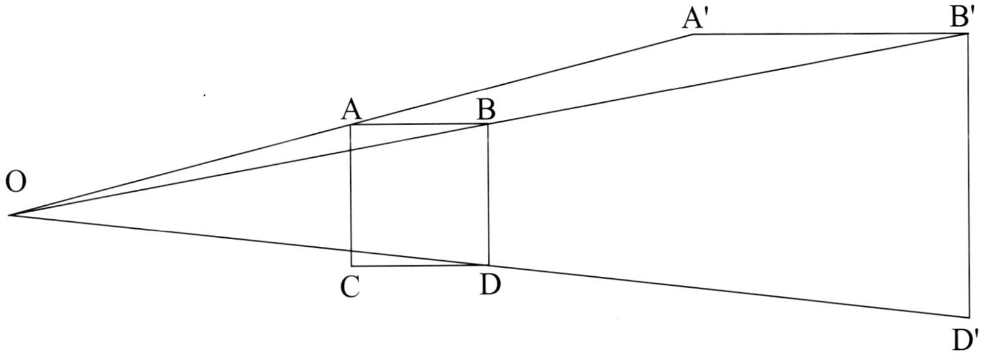

a Measure a pair of corresponding lengths, then work out the scale factor.

b Draw the diagram yourself and complete the enlargement.

c Now draw a shape of your own and enlarge it by a scale factor of 3.

Extension work

E5 Trace a map from an atlas (for example, the outline of the British Isles). Now make an accurate 'enlargement' with scale factor 1/4. What can you say about the *area* of your map compared to the original?

E6 Draw a plan of a room in your home and then enlarge it by a factor of 2.

E7 **a** Which is bigger, the circumference of your head or the length of your arm from elbow to fingertip? Write down the ratio of the two lengths.

b For about ten people, find these measurements (to the nearest cm) and then work out the ratio 'head circumference : arm length' for each of them.

c What conclusions can you draw from your results?

UNIT 3 Using percentages

A Fractions and percentages

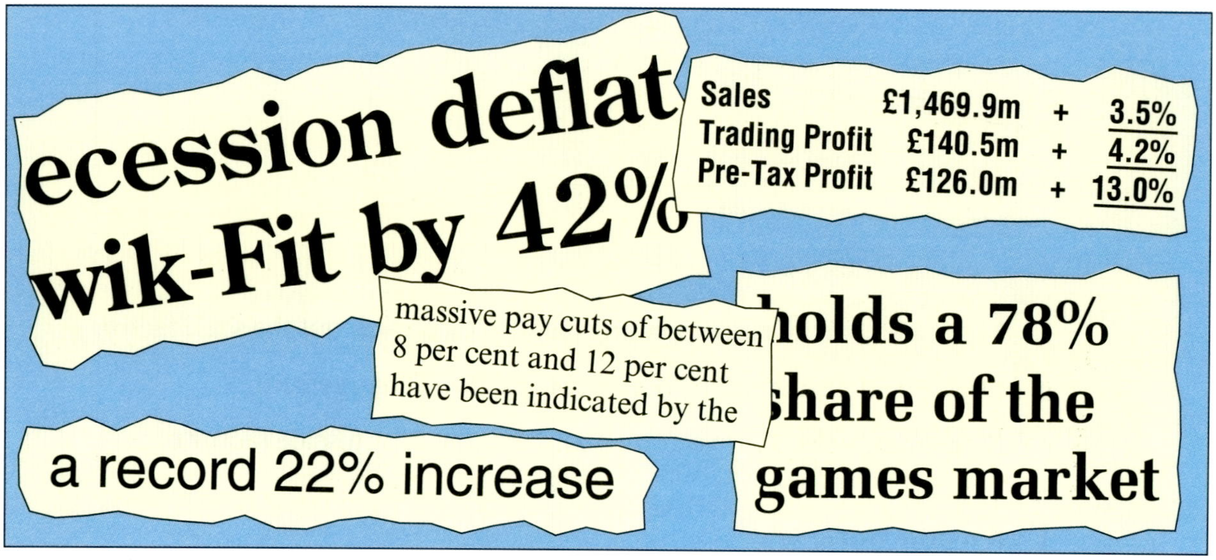

Many changes can be measured in terms of numbers.
Percentages are often used to measure the sizes of changes, but sometimes we use fractions. You have probably come across examples of the use of percentages and fractions such as:

'Salaries increase by 4%.'
'Tourism has increased by $\frac{2}{3}$.'
'Profits are down by 12%.'
'School examination passes show an increase of 6%.'
'The baby's weight increased by $\frac{1}{4}$.'
'Truancy rates are down by 1.6%.'

Look through newspapers and find as many examples as you can of the use of percentages or fractions to measure change. Why do you think we use percentages so frequently?

If school *A* had 58 passes in mathematics out of 70 entries and school *B* had 60 passes out of 100 entries, comment on the following 'headlines':

'School *B* gains more passes than School *A*'

'School *A* gains 22% more mathematics passes than School *B*'

Try to make up your own 'fair' headline.

Percentage change

To *increase* a number by 10% is the same as finding 110% of the number. This is the same as multiplying by $\frac{110}{100}$ or 1.10. For example:

To increase 500 by 10%, multiply 500 by 1.10.

$500 \times 1.10 = 550$

Similarly, to *decrease* a number by 10% means finding 90% of the number or multiplying by $\frac{90}{100}$ or 0.90. For example:

To decrease 500 by 10%, work out $500 \times 0.90 = 450$.

A1 Change these numbers as indicated:

a 180 **b** 900 **c** 50 **d** 34 **e** 2568.

i increase each by 15% **ii** decrease each by 12%

A2 A holiday firm offers 5% discount on all holidays. What would you pay for holidays priced at:

a £700 **b** £920 **c** £1040 **d** £1620?

A3 You are offered either $7\frac{1}{2}$% discount or free insurance (worth £17.50) on a holiday costing £234. Which would you choose? Why?

Fractional change

To *increase* a number by a sixth means finding 7 sixths of the number or multiplying by $\frac{7}{6}$ or 1.1666 ... For example:

To increase 300 by a sixth, multiply 300 by $\frac{7}{6}$ giving: $300 \times \frac{7}{6} = 350$.

What fraction would you need to multiply by in order to *decrease* a number by $\frac{1}{6}$?

Sometimes, you will get a lot of decimal places. For example, to increase 25 by $\frac{1}{6}$ gives $25 \times \frac{7}{6} = 29.16666666666$ ... Use your common sense to round the answer to the nearest whole number or to one decimal place.

A4 Change the numbers given in question **A1** as indicated:

i increase by a seventh **ii** decrease by an eighth.

A5 **a** Which of these sales is giving the bigger reduction?

b What would be the cost of a coat in each of these sales if it originally cost £124?

Extension work

A6 'One in 12 people in the group are over 50.'

a What percentage is this?

b If 30% of those over 50 are smokers, what fraction of the *whole* group are *both* over 50 *and* smokers?

A7 An antique dealer buys a table for £760 and sells it for £1100.

a What is the percentage profit?

b If the table had been sold for £720, what would have been the percentage loss?

(In both cases, express the change as a percentage of the original *buying* price.)

Note: questions **A8** to **A11** use the information given in question **A8**.

A8 Using the exchange rate £1 = $1.55, rewrite the following table giving the car hire rates in £ to the nearest £.

Holiday Car Hire		
Car	**Weekly**	**Additional day**
Metro	$86	$22
Cavalier	$111	$28
Chevrolet	$199	$50
Pontiac	$236	$59

A9 What would be the cost **i** in dollars **ii** in pounds of:

a a Metro for 10 days

b a Pontiac for 15 days

c a Cavalier from 5 June to 7 July inclusive?

A10 The total cost of a holiday is £1200. This includes the hire of a Chevrolet for five days.
What percentage of the holiday cost is used on car hire?
(Hint: look at your answer to question **A8**.)

A11 The cost of all car hire increases by 7%.
Make up a revised table with new prices correct to the nearest $.

A12 If unemployment increased by 2% last month and there are now 1 765 220 people out of work, how many people were out of work a month ago?

B Some unexpected percentage calculations

Extension work

B1 A secretary, who earned £250 per week, was given a 12% pay rise.

a What was the new wage?

The employer then decided that, because of falling profits, all employees must take a 12% pay cut. 'Never mind,' said the secretary, 'I'll be back on £250 a week again.'

b Is this true? (Remember: it is a 12% cut of the *new* pay rate.)

c Calculate the secretary's pay rate after the cut of 12%.

B2 In its first year after opening, a baker's shop made a profit of £2500. In the next year, the profit increased by 6%.

a What was the new profit?

b In the following year the profit increased by 8%.
'That means we have increased our profit by 14% over two years,' said the manager. Is this correct?

c Calculate the profit made in the third year.

d Calculate the percentage increase in profit over the two years.

B3 In an American store, a painting was priced at $450.

'We have to add on 7% sales tax,' said the salesperson, 'but I can give you a 5% discount.'

'Can I have the discount after you have added the tax?' said the customer.

'I thought it would be better the other way round,' said the salesperson.

Who was right? Do both calculations.

B4 'The area of the forest has declined by 1% every year for the last 100 years,' says the forester.

'That means there is no forest left at all now,' replies the listener.

a Is the listener correct? (Think carefully what 1% means. 1% of what?)

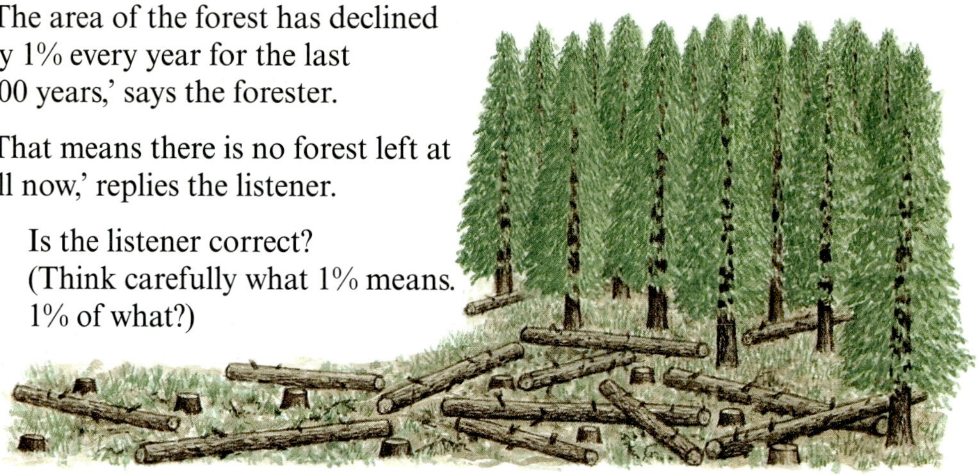

b If the area was 10 000 hectares one hundred years ago, what size will it be today?

B5 At the top of a pupil's test, the teacher had written:

'*Section A* $\frac{20}{50}$ *Section B* $\frac{30}{50}$

$\frac{20}{50} + \frac{30}{50} = \frac{50}{100} = 50\%$'

'That can't be right,' said the pupil, '$\frac{20}{50} + \frac{30}{50} = \frac{50}{50} = 100\%$. I should get full marks!' Explain.

C Apple picking

Extension work

C1 An apple picker picks 20 kg on Sunday.
On Monday, the total picked is half as much again as on Sunday.
On Tuesday, the total is a third as much again as Monday.
On Wednesday, the total is a quarter as much again as Tuesday.
On Thursday, the total is a fifth as much again as Wednesday.
On Friday, the total is a sixth as much again as Thursday.

a What weight of apples was picked on Friday?

b If the picking continued with the same pattern of increases, what weight of apples would be picked on:

i the 20th day **ii** the 100th day?

C2 Another apple picker also picks 20 kg on Sunday.

On each day after Sunday, the weight picked is 10% more than the previous day.

What was the weight of apples picked on:

a Tuesday **b** Thursday **c** Saturday?

D Practice with percentages

You often need to write one number as a percentage of another. For example, write 6 as a percentage of 150.

First find 6 as a decimal fraction of 150:

$\frac{6}{150} = 6 \div 150 = 0.04$

Then multiply by 100 to find the percentage equal to 0.04:

$0.04 \times 100 = 4$ (so $0.04 = \frac{4}{100} = 4\%$)

So 6 is 4% of 150.

D1 Express, correct to two decimal places:

a 8 as a percentage of 150 **b** 17 as a percentage of 289

c 0.32 as a percentage of 1 **d** 0.32 as a percentage of 10

e 0.32 as a percentage of 100.

D2 Reduce each of the following by 12%:

a 100 **b** 1000 **c** 5000 **d** 162 **e** 17.4 **f** 0.001 **g** 0.73.

D3 Increase the numbers in question **D2** by $7\frac{1}{2}\%$.

D4 A rectangle has sides of length 3 cm and 4 cm.

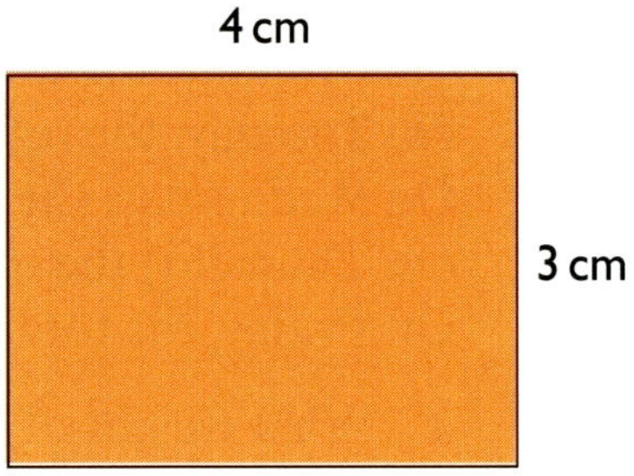

a What is its area?

b If each side is increased by 50%, what is its new area?

c What is the percentage increase in area?

D5 Here is a mixture of fractions, decimals and percentages:

a 18% **b** $\frac{1}{3}$ **c** 0.72 **d** 85% **e** $\frac{2}{7}$ **f** 0.44 **g** 57% **h** $\frac{3}{8}$ **i** 0.59

i Write them all as decimals.

ii Put them in order from the smallest to the biggest.

E Supermarket sales

Extension work

The table shows the amounts of money taken by two branches of a supermarket during one week.

Day	Money taken Branch A	Branch B
Monday	£3 756	£5 476
Tuesday	£7 643	£8 322
Wednesday	£8 897	£10 961
Thursday	£12 854	£13 682
Friday	£18 453	£21 224
Saturday	£22 187	£25 666

E1 Write a short report about the amounts of money taken by the two branches. Use percentages whenever possible.

You might comment on, for example:

What percentage of the money is taken on Saturday?

Which branch shows the bigger percentage increase from Monday to Saturday?

How do the amounts taken at the two branches compare?

Draw a graph to illustrate your comments where possible.

F Percentages and maps

Extension work

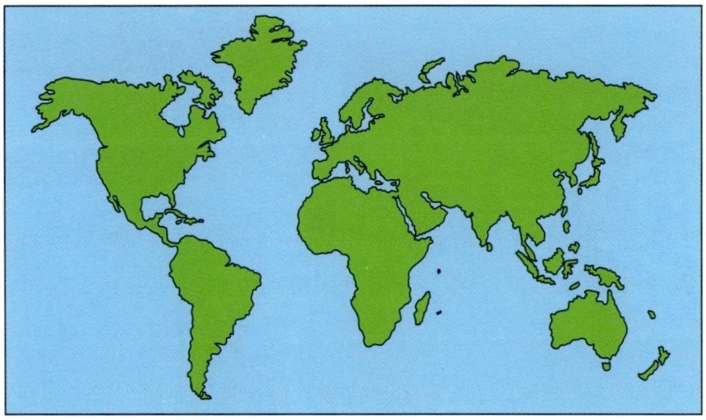

Map of the World

N3:1

F1 Obtain one of the following (or all three if you can):

a map of the world

a plan of your nearest town or district

a plan of your school.

Find the percentage of different types of space on at least one map.

(These questions are difficult to answer accurately: because of the curved surface of the earth, maps often distort area. Give rough approximations for the map or plan that you have.)

For example, on a map of the world:

What percentage of the Earth's surface is water?

What percentage of the Earth's land is north of the equator?

What percentage of the Earth's land is in Africa?

For example, for your town map:

What percentage of the town consists of parks/open spaces?

What percentage of the town consists of shops?

For example, for your school plan:

What percentage is buildings?

What percentage is playing fields or playground?

What percentage is occupied by car parks?

G A cross-number puzzle

N3:2

G1 Make a copy of this grid or use Worksheet N3:2.

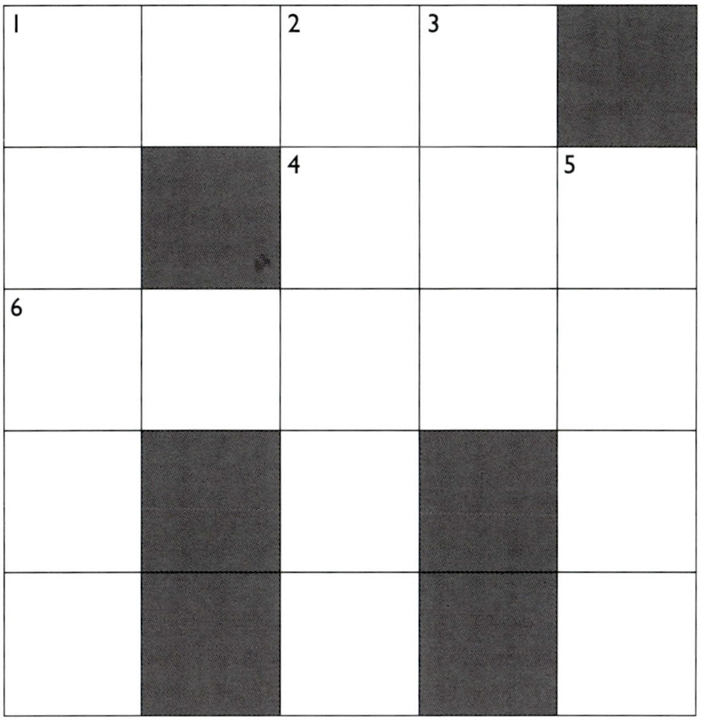

Use the clues to fill in your cross-number.

Across

1. 75% of 9852

4. Decrease 699.6 by a sixth

6. Increase 28 232.4 by two thirds

Down

1. Increase 60 000 by 19%

2. Decrease 100 000 by 15%

3. $12\frac{1}{2}$% of 7880

5. 20% of this number is 683.4

UNIT 4 Number patterns

A Exploring a number square

1	2	3	4	5	6	7	8	9	10
11	12	13	14	15	16	17	18	19	20
21	22	23	24	25	26	27	28	29	30
31	32	33	34	35	36	37	38	39	40
41	42	43	44	45	46	47	48	49	50
51	52	53	54	55	56	57	58	59	60
61	62	63	64	65	66	67	68	69	70
71	72	73	74	75	76	77	78	79	80
81	82	83	84	85	86	87	88	89	90
91	92	93	94	95	96	97	98	99	100

N3:3

You will need a copy of Worksheet N3:3.

Start at any number in the square, 13 for example, and move 2 squares to the right and 1 down. You reach the number 25. Repeat the process from 25 to reach 37. This is the start of a sequence of numbers, which we could call the (2 right, 1 down) or (2R, 1D) sequence.

(2R, 1D) is called the **sequence generator**. You can record the sequence and its name like this:

(2R, 1D) → 13, 25, 37…

Continue the process from 37. How can you continue when you reach the edge of the square or the bottom?

A1 Describe the pattern in this sequence of numbers.

A2 Start in another square and write down the sequence you get from (2R, 1D). Does it produce a similar pattern? Check with other starting points.

A3 Choose your own starting points and write down the sequences you get from:

a (2R, 1D) **b** (1R, 2D) **c** (3R, 3D)

d (0R, 2D) **e** (5R, 1D).

A4 Describe the sequences for each of the generators in question **A3**. Make a table of your findings like this:

Generator	Description of sequence
1R, 1D 1R, 2D	

A5 What is the connection between the generator and the sequence?

A6 Predict what sequences you would get from these generators:

a (2R, 4D) **b** (0R, 5D) **c** (10R, 10D) **d** (pR, qD).

A7 L stands for 'left' and U for 'up'. What sequence would you expect from the generator (2L, 3U)? Check it out on the number square.

A8 What would be the generators for these sequences:

a 17 28 39 50 61 …

b 3 12 21 30 …

c 84 71 58 45 …

d 150 132 114 96 …?

A9 The generators (4R, 2D) and (6L, 3D) produce the same sequence.

a Explain why.

b Find two other generators which produce the same sequence as each other.

B Number puzzles

B1 Repeater

Think of a 3-digit number (such as 245).
↓
Multiply it by 7.
↓
Multiply your answer by 11.
↓
Multiply your last answer by 13.
↓
What do you notice about the final answer?

B2 **a** What happens if you use 2-digit numbers in question **B1**?

b What is $7 \times 11 \times 13$ anyway?

c Try question **B1** with different 3-digit numbers. Does the same thing always happen?

d Can you explain why you get these answers?

B3 What would you have to multiply a 4-digit number by to get the same effect as in questions **B1** and **B2**?

B4 Endless loop

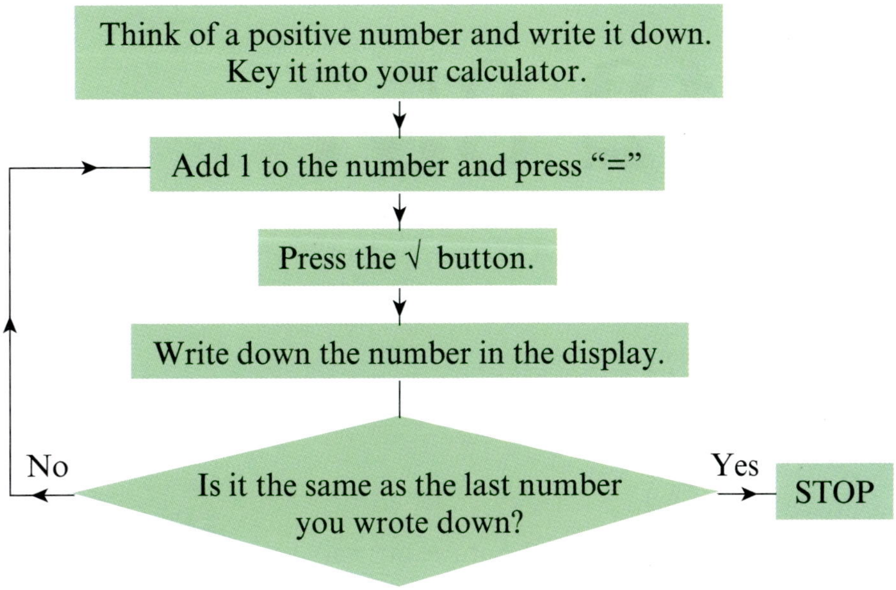

a Try several different numbers in this flow chart and record your results. What pattern did you find?

b Is there a pattern when you start with:

i 0 **ii** 1000 **iii** 1 000 000?

c Is there a pattern when you start with negative numbers?

d What connection does this have with the *Fibonacci Series*?

C Cyclic numbers

142 857

Cyclic means ‘related to circles’. To see why this number is called cyclic, complete the following multiplications:

142 857 × 2 = 285 714
142 857 × 3 =
142 857 × 4 =
142 857 × 5 =

C1 What do you notice about the answers you get compared with the original number 142 857?

C2 Without doing any multiplication, predict the answers to:

a 142 857 × 6 **b** 142 857 × 7

Now check your answers.

You can think of the six figures 1 4 2 8 5 7 as being in a circular chain. This chain can be broken at any one of six points to produce the first six multiples of 142 857. This is why 142 857 is described as cyclic.

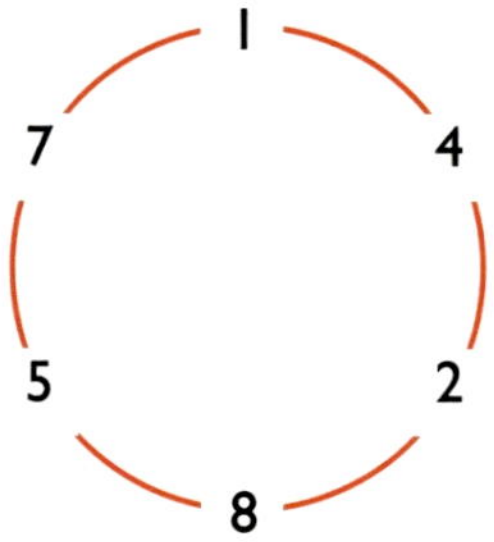

The number 142 857 occurs in many places in mathematics. You may have come across it changing $\frac{1}{7}$ to a decimal. If you work out 1 ÷ 7 on a calculator you get the recurring decimal:

$\frac{1}{7}$ = 0.142 857 142 857 14…

(If your calculator can only display eight decimal places, use LOGO which gives 15.)

C3 **a** What *exactly* do you get on your calculator if you multiply this recurring decimal by 7?

b Does this explain why 142 857 stopped being cyclic when you multiplied it by 7 in question **C2 b**? Explain your answer.

Another fraction that produces a recurring decimal with cyclic properties is $\frac{1}{13}$. Check on your calculator that:

$\frac{1}{13} = 0.\mathbf{076\,923}\,076\,9\ldots$

C4 **a** Investigate the cyclic properties of the number 076 923 by completing the following multiplication table:

076 923 × 1 = 076 923
076 923 × 2 = 153 846
076 923 × 3 =
076 923 × 4 =
076 923 × 5 =
076 923 × 6 =
...
076 923 × 12 =

b How many cycles has this number got? Should we call it a bicyclic number?

c When you have completed your multiplication table for 076 923, make a list of all the patterns you can see in the numbers. Here are a few ideas to think about:

- look at the first figure in each number in the table
- look at the last two figures in each number
- compare the last number in the table with the first; also try comparing the next to last with the second.

C5 Compare the three cyclic numbers you have found:

142 857 076 923 153 846

- add up the digits of each number
- break each 6-figure number into two 3-figure numbers and add them together like this:

  142
  +857

- try some other ways of comparing the numbers.

D A number spiral

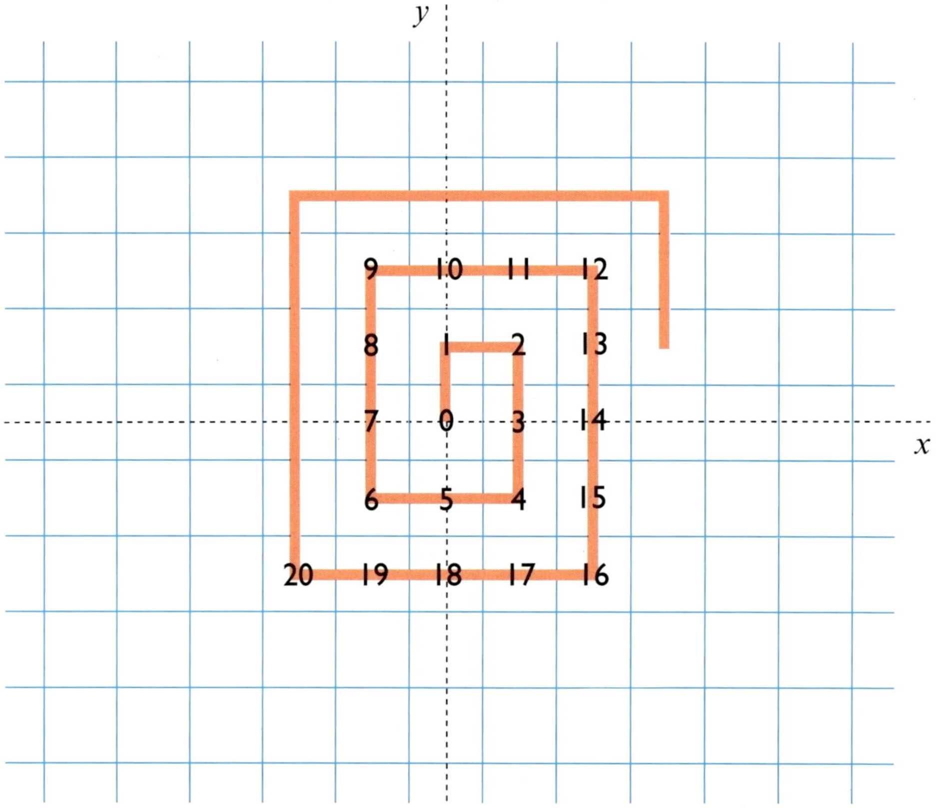

Copy this spiral of numbers on to a sheet of squared paper and continue it at least as far as 50.

D1 Are there any patterns in the way certain groups of numbers are distributed on the grid? For example, how are *odd* and *even* numbers spread out?

The positions of numbers on the grid can be described using x- and y-axes and counting squares horizontally and vertically from 0. Using this method 13 is at position (2, 1) and 17 is at (1,–2).

D2 Find the square numbers. Describe their position in the grid.

D3 The numbers 4, 16, 36 … are at (1, –1), (2, –2), (3, –3) …

a Without filling in all the numbers, find out what the positions of 64, 100 and 144 will be.

b If n is an even number where will n^2 be?

c What are the positions of the square numbers 1, 9, 25, and 49?

d Without filling in all the numbers, say what the positions of 81, 121 and 169 will be.

e If m is an odd number, where will m^2 be?

You can make a sequence of numbers by multiplying consecutive pairs of whole numbers together:

$1 \times 2 \quad 2 \times 3 \quad 3 \times 4 \quad 4 \times 5 \quad 5 \times 6 \ldots$

D4 Work out the numbers in this sequence and write down their positions in the grid.

a Is there a pattern to these positions?

b Is the position affected by whether the smaller number is even or odd?

c Try to give a general position for $p \times q$ where p and q stand for consecutive numbers.

D5 Make a table of values to find the sequence of numbers produced by the expression $4x^2 - x$. Choose different values of x such as 1, 2, 3, 4 and complete this table.

x	1	2	3	4
x^2	1	4		
$4x^2$	4	16	…	
$-x$	−1	−2	…	
$4x^2-x$	3	14	…	

a Where can you find this sequence on the grid?

b What sequence do you get if you put negative values of x into the expression $4x^2 - x$? (For example if $x = -2$ you get the value 18 for the expression.) Find this sequence on the grid.

D6 **a** Find the sequence of numbers in the grid that runs:

71 41 19 5 3 13 31 57

b Evaluate the expression $(4x^2 - 2x + 1)$ for values of x such as 1 2 3 4 … This will produce half of the sequence for you. Try to find an expression that will produce the other half.

E Number squares revisited

This number square has only six columns, labelled A to F. It also has six rows, but you could extend it downwards to numbers greater than 36.

A	B	C	D	E	F
1	2	3	4	5	6
7	8	9	10	11	12
13	14	15	16	17	18
19	20	21	22	23	24
25	26	27	28	29	30
31	32	33	34	35	36

E1 **a** Choose a number from column B and a number from column C. Add them together. Which column has the answer in it?

b Choose two more numbers from B and C and add them together.

Find which column contains the total.

c Find the result of adding numbers from columns C and D together.

d Make an addition table like the one below. This table shows where you would find the answer when adding numbers from different columns together.

+	A	B	C	D	E	F
A						
B			E			
C		E				
D						
E						
F						

e Try to explain why adding a number from any column to a number from column F always produces an answer which is in the original column.
For example:

B + F = B; C + F = C.

(Look back to the 10-column number square on page 31 of this Unit. Which column has an effect similar to F?)

E2 **a** Look again at the number square above. Choose a number from column B and multiply it by any number from column E. Which column contains the answer? Check that the same connection holds for other numbers in columns B and E (extend the columns if you need to).

b Make a multiplication table like the one below to show the results of multiplying numbers from different columns together.

c What is the effect of multiplying anything by a number from column F?

d What is the effect of multiplying anything by a number from column A?

e Try to explain why the *diagonal of the addition table* is the same as the *B column of the multiplication table.*

×	A	B	C	D	E	F
A						
B					D	
C						
D						
E		D				
F						

E3 If you were unable to answer question **E2 e** try the following:

a Go back to the original 6-column number square. Divide each number by 6 and write down the *remainders* in a table like this:

A	B	C	D	E	F
1	2	3			
1	2				
1					

You should find that the A column is full of 1s, B is full of 2s and so on.

b Replace the letters in the addition and multiplication tables in questions **E1** and **E2** with the matching numbers:

A 1 B 2 C 3 D 4 E 5 F 0.

c Do they still work as addition and multiplication tables?

d Compare the diagonal of the addition table with the B column of the multiplication table. What does each represent? Why must they be the same?

UNIT 5 Sequences

A Making sequences

A1 Find the next two terms in each of these sequences:

a	1	7	13	19	...	...
b	3	9	27	81	...	...
c	7	12	17	22	...	...
d	−10	−6	−2	+2	...	...
e	12	5	−2	−9	...	...
f	30	41	52	63	...	...
g	100	81	62	43	...	...

A2 Find out how your calculator does repeated addition.

a Use your calculator in this way to produce the sequences of numbers in question **A1** parts **a** and **c**.

b Continue with your calculator to find the 20th and 25th terms in the sequences.

A3 **a** Find the sequence of keys which produces each of the other sequences in question **A1**.

b Find the 19th term of each sequence.

A4 **a** Find the key strokes on your calculator which produce the sequence that begins:

9 17 25 33 41 ...

b How many terms in the sequence are needed before it passes 500?

c Which term in the sequence is closest to 250?

A5 **a** Use your calculator to produce the sequence:

149 138.5 128 117.5 ...

b Which is the first term less than zero?

c How many terms are there greater than −100?

B Using a formula

You can make a sequence by taking a formula such as

$3n + 2$

and putting different values of n into it.
Putting $n = 1, 2, 3 \ldots$ gives:

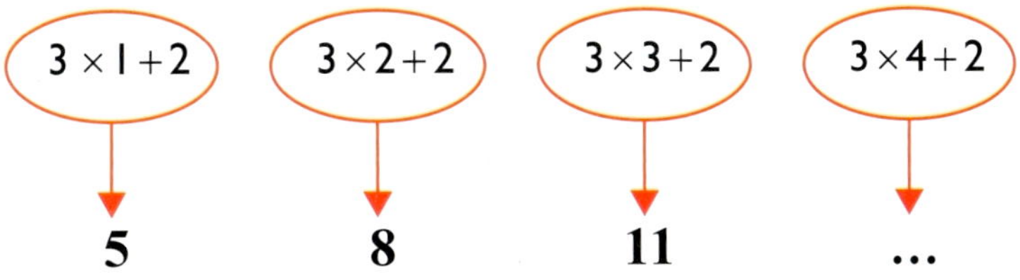

B1 Write down the first six terms of the sequence you get from the formula $3n + 2$.

B2 Write down the first six terms of the sequences generated by:

a $2n + 5$ **b** $3n - 6$ **c** $n^2 + 8$

d $3 - n$ **e** $10 - 5n$

B3 What calculator key strokes will produce the sequences in question **B2**?

B4 **a** Write down the sequence you get from the formula $4n - 7$ for values of n from 1 to 6.

b Find the 100th term of this sequence.

c How many terms would you have to go through before reaching one that is bigger than 200?

B5 This program in BASIC will produce the first 20 numbers of the sequence that comes from the formula $4n + 3$:

```
10  FOR N = 1 TO 20
20  PRINT N, 4*N + 3
30  NEXT N
40  END
```

a Run this program and read off the 13th and 17th terms of the sequence.

b Modify the program to produce the first 4000 terms of the sequence. Run it.

c Which is the first term in the sequence which is greater than 7000? (You may need to use the 'escape' key while the program is running.)

B6 The following program will produce the sequence of up to 500 numbers which comes from the formula $6n - 11$ but it will stop when it reaches a number greater than 1000:

```
10  FOR N=1 TO 500
20  PRINT N, 6*N-11
30  IF 6*N-11>1000 THEN 50
40  NEXT N
50  END
```

Run the program and find the first term in the sequence which is greater than 1000.

B7 **a** Write a program which will precede the first 2500 terms of the sequence that comes from the formula $1000 - 13n$.

b Read off the 1250th term and the 1750th term.

c How many terms are greater than –2000?

d Modify the program to stop once the terms in the sequence become negative.

B8 Ampshire Electricity Company charges customers according to the formula

$$£9.86 + 7.59\text{p} \times N$$

where £9.86 is the fixed standing charge, 7.59p is the cost of a unit of electricity and N is the number of units used. Write a program to show the charges for supplying up to 2000 units of electricity.

C Problems

C1 An ornamental garden fence is made of pillars and cross-beams as shown. The pillars are placed 1 m apart.

a Make a table showing the number of pillars and cross-beams (there are four between each pair of pillars) needed for 1 2 3 … up to 10 m of fencing.

b Predict how many pillars and cross-beams would be needed for 50 m and 100 m of fencing.

c Write in words, or as a formula, the connection between:

i the length of fencing and the number of pillars

ii the number of pillars and the number of cross-beams.

C2 An ancient jewellery design involves placing emeralds at the vertices of gold pentagons. These diagrams show the settings for one, two and three pentagons.

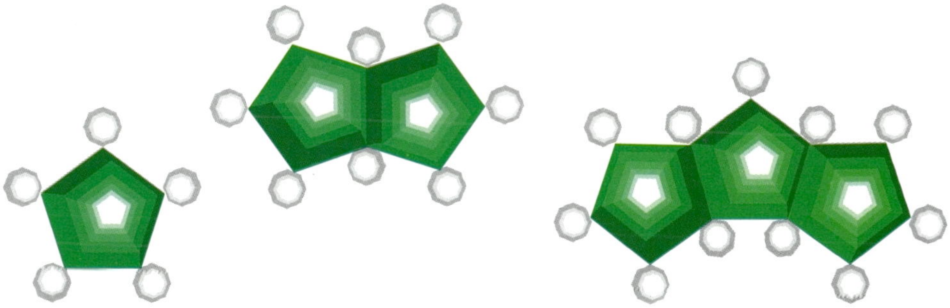

a How many emeralds would be needed for four and for five pentagons?

b Predict how many would be needed for a chain of 12 pentagons.

c Suggest a formula that connects the number of emeralds with the number of pentagons.

C3 A café owner decides to try out some triangular tables. Around one table on its own she can seat three customers. Around two tables (joined edge to edge) she can seat four.

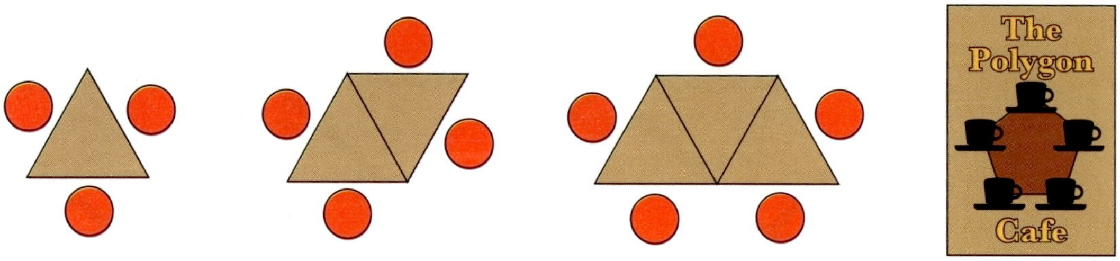

a How many customers can be seated round four, five or six tables (joined edge to edge) in a line?

b Suggest a formula that connects the number of tables (joined together) with the number of seats.

In a survey of their preferences, the customers say that the single table is too crowded. Most people prefer two or three tables joined together. The owner has 60 tables available. She puts some of them in pairs and some in threes and finds she has created seats for 110 customers.

c How many tables are in pairs and how many are in threes?

C4 You can make a paper chain by pasting strips of paper together. Some children make one from strips of paper which are 30 cm long. They paste them together so that 5 cm of each strip overlaps at each joint.

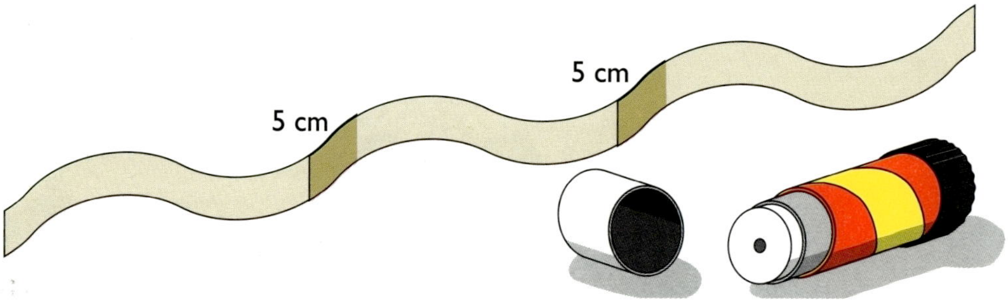

a Make a table to show the lengths of chain made by one two three … up to ten strips.

b Would 20 strips make a chain twice as long as ten strips?

c Predict how long a chain of 25 strips would be.

d How many strips would be needed to make a chain at least 5 m long?

e Suggest a formula which will enable you to work out how long any number of strips would be.

C5 A pile of five stacked chairs is 120 cm high.
Six stacked chairs make a pile 132 cm high.

a How tall will a pile of ten stacked chairs be?

b How tall is one chair?

C6 A plastic storage tray is designed with a 'hook and lip' so that it can be interlocked with others. Ten trays linked together have a total width of 114 cm. Twelve trays have a width of 136 cm.

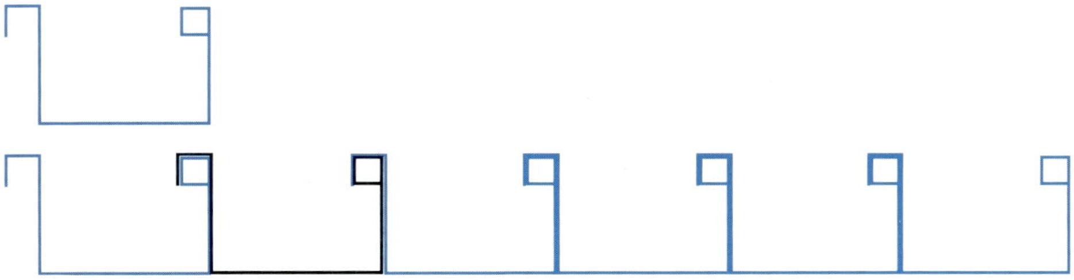

a How wide will 16 trays be?

b How wide is one tray?

C7 A pattern design consists of thick and thin stripes. The pattern always begins and finishes with thick stripes.

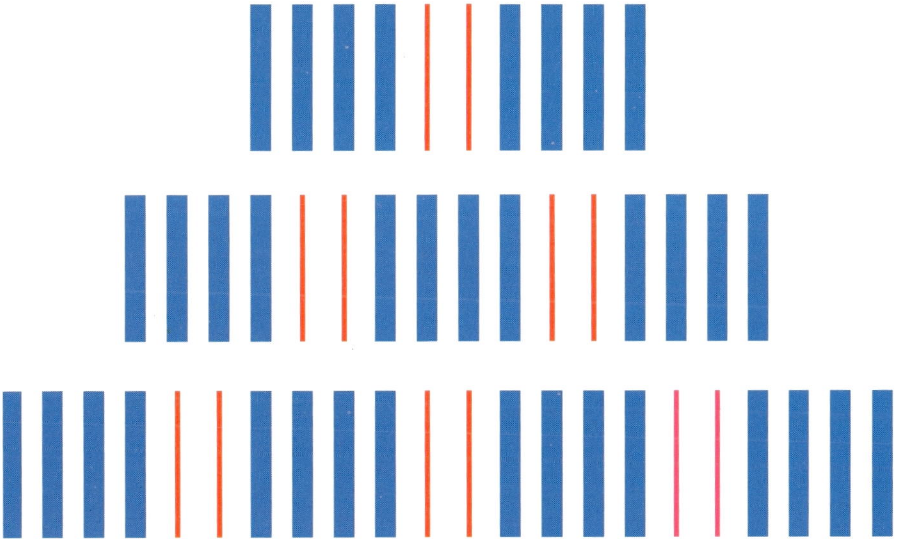

a Make a table to show the number of thin stripes and the number of thick stripes that appear in various widths of the pattern.

b Suggest a formula which connects the number of thin stripes (n) to the number of thick ones (k).

D Sequences using LOGO

LOGO

D1 **a** Make up three sequences which begin with:
1, 3 ...

b Write down the tenth and 21st terms for each of your sequences.

D2 Make up sequences which begin with:

a 2, 4 **b** 5, 7 **c** 30, 100 **d** 35, 30.

i Describe how you move from term to term.
ii Predict the tenth term.
iii What would the 21st term be?

D3 What sequences are produced by these LOGO procedures?

a
```
to addfour :n
print :n
addfour :n+4
end
```

b
```
to takefour :n
print :n
takefour :n-4
end
```

c
```
to addten :n
print :n
addten :n+10
end
```

b
```
to taketwo :n
print :n
taketwo :n-2
end
```

D4 By using different start numbers and rules, design sequences with:

a 10 as the sixth term **b** –3 as the seventh term

c 1.5 as the third term **d** 12 as the 12th term

e Write LOGO procedures which will print the sequences you devised for parts **b** and **c**.

E Using spreadsheets

Spreadsheet

The tables below show spreadsheets where column A gives the term of the sequence and columns B and C are generated by the formulae shown.

	A	B	C
1	Term	Seq	Seq
2	1	7	7
3	2	12	12
4	3	17	17
5	4	22	22
6	5	27	27
7	6	32	32

	A	B	C
1	Term	Seq 1	Seq 2
2	=1	=7	=5*A2+2
3	=A2+1	=B2+5	=5*A3+2
4	=A3+1	=B3+5	=5*A4+2
5	=A4+1	=B4+5	=5*A5+2
6	=A5+1	=B5+5	=5*A6+2
7	=A6+1	=B6+5	=5*A7+2

E1 **a** Set this up on a spreadsheet so that the first ten terms can be seen.

b Change the number in cell B2 (in the spreadsheet it starts as a 7).

c Change the formula in column C.

d Try other numbers in cell B2.

E2 Repeat question **E1** for these sequences:

a 50 43 36 29

b 5 6 7 8 9

c 10 8 6 4 2.

d For each sequence answer these questions:

What is the tenth term of your sequence?
What is the 100th term?
What is the 99 000th term?

F Sequences in sequences

Design a sequence which begins 3, 7

3, 7, 11, 15, ... just add 4

That's nice! A sequence within a sequence

How about 3, 7, 12, 19, ... add 4, then 5, then 6

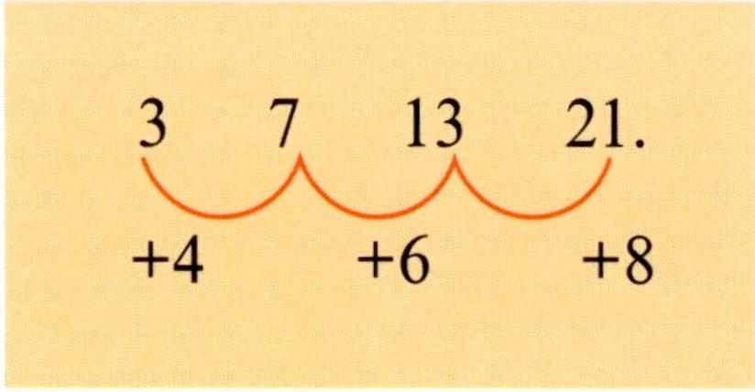

I can draw one like that.

I can set that up on a spreadsheet

	A	B
1	Hidden	
2	Sequence	Sequence
3		3
4	4	7
5	6	13
6	8	21
7	10	31

	A	B
1	Hidden	
2	Sequence	Sequence
3		3
4	4	B3+A4
5	=A4+2	=B4+A5
6	=A5+2	=B5+A6
7	=A6+2	=B6+A7

F1 In a small group, discuss what is happening in the picture above. Set a starting point of your own and draw your own picture to describe what you are doing.

Spreadsheet

F2 In the following, find the sequence within the sequence. Set up each sequence on a spreadsheet, as in the picture. Remember to make notes of all your working so that you can compare your results with someone else's. Work on as many of these as you have time for.

a 2 5 10 17 26 37 50 65 82 …

b 6 9 14 21 30 41 54 69 86 …

c −10 −7 −2 5 14 25 38 53 70 …

d 2 3 8 17 30 47 68 93 122 …

e 7 8 13 22 35 52 73 98 127 …

f 1 3 6 10 15 21 28 36 45 …

g 6 8 11 15 20 26 33 41 50 …

h 1 21 40 58 75 91 106 120 133 …

i –40 –20 –1 17 34 50 65 79 92 …

j 1 5 14 30 55 91 140 204 …

k 1 4 10 20 35 56 84 120 …

l 1 4 9 16 25 36 49 64 …

m 1 9 30 70 135 231 364 540 …

F3 Do any of the sequences in question **F2** have hidden sequences in common? If so, write down these sequences.

F4 Design some sequences within sequences on a spreadsheet. For each sequence find the 15th term. Extend your sequence in the negative direction and find the –10th term.

F5 Devise some sequences within sequences for each other. In each case find the 20th and the –20th terms.

F6 Find as many different sequences as you can which begin:

2 4 8 …

Which of them can be designed using hidden sequences on a spreadsheet? Set them up on a spreadsheet if possible.

F7 For each of the following find three different sequences which have the given numbers as the first two terms:

a 1 5 …

b 1 2 …

c 9 18 …

d –6 6 …

e 0.5 0.8 …

f 102 106 …

g 17 51 …

h 64 8 …

UNIT 6 Numbers in everyday life

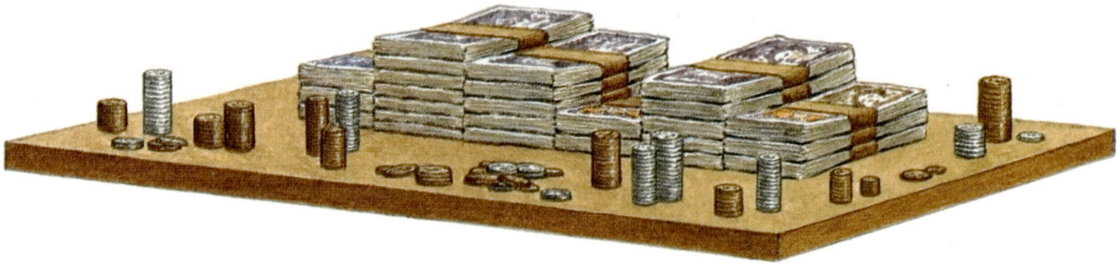

Investing and borrowing money

One of the most frequent uses of percentages is in calculating interest on money borrowed or invested. If you *borrow* money, for example, to pay for a house or a car, you have to pay back both the sum borrowed and an additional amount of *interest*. If you have money to spare and deposit it in a bank or building society, then you normally *receive* interest on your money.

A Investing money

If you invest £500 in a building society at a rate of interest of 6% per annum, this means your money will earn interest of $\frac{6}{100} \times £500$ in a year.

At the end of the year you will have $0.06 \times £500 = £30$ in interest. When this is added to your original £500, it makes £530 altogether.

A1 What interest do you receive if you invest £700 for one year at the following rates of interest? Give each answer to the nearest penny.

a 6% **b** 7% **c** 11% **d** $5\frac{1}{2}$% **e** 3%

f $2\frac{1}{2}$% **g** 12% **h** 3.14% **i** 2.15% **j** 8.25%

A2 If you do not withdraw any money, but instead keep it in the building society for another year, you earn interest on the previous year's interest. For example, if you invest £100 at 6% after one year you will have £106.

In the second year, you will have interest of $\frac{6}{100} \times £106$.

How much will you have altogether after two years?

(Note: this is called getting **compound interest** on the money invested.)

A3 If the interest rate is 7% per annum compound interest, what will the following amounts be worth after two years?

a £100 **b** £200 **c** £1000 **d** £100 000 **e** £8000 **f** £1m.

For each year, you multiply the amount you have at the beginning of the year by 0.07 (since the interest rate is 7% per annum) to find the interest for that year.

So, starting with £100, you have £100 × 1.07 = £107 at the end of year one.

For the second year, you multiply £107 × 1.07 = £114.49.
The amount after two years is obviously the same as
£100 × 1.07 × 1.07 = £114.49.

The amount after three years would be
£100 × 1.07 × 1.07 × 1.07 = £122.5043 = £122.50 (to the nearest penny).

This method saves time if you have several years of interest to calculate.

A4 If you invest £1000 at an interest rate of 6.5%, what will your investment be worth after:

a two years **b** five years **c** ten years?

A5 If you invest £100, how many years will it take you to double your money at interest rates of:

a 10% **b** 5%?

You might like to guess the answer first and then see whether you are right.

Extension work

A6 Sometimes, interest is paid monthly. So, if you invest £100 at 6% per annum, your account will be worth £100 + $\frac{1}{12}$ × £6 at the end of the first month. You then receive interest on the new amount for the next month, and so on.

What will your account be worth at the end of:

a the second month **b** the fifth month **c** the first year?

Building societies advertise a 'CAR' for a monthly interest account. Find out what CAR means and calculate it for the example above.

A7 Imagine that you have £10 000 to invest. Look at advertisements in newspapers, or look in the windows of building societies, to decide which one would give you the best rate of interest.

When you look at interest rates, you will find that they are often quoted as 'gross' and 'net'. The 'net' rate is after the deduction of tax. Use the 'gross' rate for the calculations below, assuming you are a non tax-payer.

How much would your money be worth after one year in:

a the best account you can find **b** the worst account?

A8 Your local Post Office has leaflets on 'National Savings', which is money lent to the government, usually for a minimum length of time.

Imagine that you have £5000 to invest in National Savings. What particular form of savings would you choose and how much would your money be worth after five years?

A9 Compare the rates of interest offered by building societies and National Savings.

Make lists of advantages and disadvantages of investing money in either of these.

B Borrowing money

Hire purchase

Big items like motorbikes or CD players can often be bought on hire purchase. People buy an item by paying a deposit which is part of the purchase price. Then they repay the remainder in instalments over a period of months – usually 12, 24 or 36. Interest is paid on the amount owed. For example:

A mountain bike costs £200. A deposit of 10% is required. The rate of interest for a 12 month loan is 15%. Calculate the deposit and the 12 monthly payments.

The deposit is $0.10 \times £200 = £20$, which leaves £180 to pay.
The interest on £180 at 15% for 12 months is $0.15 \times £180 = £27$.
So the total to be repaid is $£180 + £27 = £207$.
The monthly repayments are $£207 \div 12 = £17.25$.

B1 Fatima bought a computer costing £400. She paid a deposit of £40 and agreed to pay the remainder of the money in 12 monthly instalments.

a How much would she pay each month if no interest were charged?

b If interest were charged at 10%, how much interest would she pay altogether? How much is this per month?

c What would be the monthly payments to pay off both the sum owing and the interest?

B2 Calculate the monthly payments over 12 months on the following:

a Television set: price £300, deposit £40, interest rate 10%

b Bicycle: price £250, deposit £50, interest rate 15%

c Car: price £8000, deposit £800, interest rate 5%.

Extension work

B3 Collect examples from newspapers, local shops, garages, etc. of items being offered for sale where a deposit is needed followed by a number of monthly payments.

For six different examples, calculate the difference between the actual cost of the item and the amount you would pay using a deposit and instalments.

When you collect your examples, you may find that an 'APR' percentage rate is quoted. Find out what it means. This will show more clearly why interest rates for hire purchase may not be what they seem.

Mortgages

If you buy a house, it is unlikely that you will have enough money to pay for it in full. Normally you pay a deposit and borrow the rest of the money from a bank or building society.

A loan like this is called a **mortgage**.

The banks and building societies set rates of interest which you must pay on the money borrowed. These rates are called **mortgage rates**.

B4 A house is bought for £40 000. A deposit of £10 000 is paid and the remainder paid by monthly instalments.

Over 15 years, the cost would be £247.20 per month.
Over 20 years, the cost would be £213.15 per month.
Over 25 years, the cost would be £198.75 per month.

For example:
Over 15 years, the repayment will be £247.20 each month.
That is 12 × £247.20 = £2966.40 each year.
That is 15 × £2996.40 = £44 496 over 15 years. The total amount paid is the deposit of £10 000 plus £44 496 = £54 496.

a Calculate the *total* amount (including the deposit) that would be paid over 20 years and over 25 years.

b If, instead of buying the house, you paid a rent of £182 a month for 15 years, how much rent would you pay altogether?

Extension work

B5 Find out, from your local newspaper or estate agents, the prices of five houses for sale near where you live.

For each house, work out the total cost over 15 years if:

- a deposit of £10 000 is paid
- monthly repayments are £90 for every £10 000 borrowed.

C A number quiz

You will need to look at the page opposite to answer the following questions.

C1 For how many hours is the library open in a full week?

C2 How many ice hockey games have been played so far?

C3 What will the bill be if I buy 2lb of each of the three cheeses?

C4 What is the cost per ball if I buy 80 and include p+p in the overall price (to the nearest p)?

C5 What is the percentage saving if I buy nine 'wonder' windows instead of nine ordinary ones?

C6 The library plans to close at 17.00 in future on Monday, Tuesday, Thursday and Friday and at 11.30 on Wednesday and Saturday. What will be the percentage reduction in opening hours?

C7 I buy $3\frac{1}{2}$ lb Cheddar and $14\frac{1}{2}$ lb Gloucester. What is my bill?

C8 What percentage of ice hockey games played so far have been draws?

C9 I buy seven ordinary windows and 15 'wonder' windows. I am given a 7% discount. What is my bill?

C10 I wish to buy *exactly* 150 golf balls. What is the total cost using the cheapest way possible?

C11 The library assistant is paid £4.32 per hour. What is the assistant paid for 4 weeks' work?

C12 I wish to spend £100 and buy equal amounts of each cheese. How many lbs (to the nearest lb) of each will I get?

ICE HOCKEY LEAGUE

	Played	Won	Drawn	Lost	Points
Tigers	6	3	2	1	8
Lions	5	3	0	2	6
Panthers	6	2	1	3	5
Leopards	5	1	1	3	3

D Estimating and checking

D1 George and Alice Brown needed a new carpet. They measured their room and found it was 5.4 m long and 3.2 m wide. With their calculator, they worked out the area of the carpet to be 172.8 m^2. Is this correct? Use an estimate to explain why.

If you always **estimate** the size of an answer, you should avoid making silly mistakes.

D2 What would be a sensible estimate for 27.8×0.3?

Did you suggest about a third of 27?

D3 Estimate, without using a calculator, the following:

a 27×9 **b** 27.8×6 **c** 0.278×15 **d** $27.8 \div 3$

e $278 \div 39$ **f** 0.7×0.63 **g** 83×27 **h** 0.8734×54

i 0.75894×1009 **j** 0.45×0.00045.

Now use your calculator to work out the exact answers.
If your estimates have been good, you should spot errors very quickly when using your calculator!

D4 Estimate the total of all the ages of the pupils in your class.

D5 Estimate the area of your classroom.

D6 Estimate how long it would take if you travelled from your school to the centre of London in a car averaging **a** 20 **b** 30 **c** 40 miles per hour.

D7 Estimate the number of separate panes of glass in:

a your home **b** your school.

D8 Estimate the following:

a the weight of a pair of shoes

b the height of a door in your classroom

c the capacity of a bucket

d the number of blades of grass in 1 m^2 (try a small area, such as 1 cm^2, and multiply by 10 000).

Check, where you can, by measuring.

D9 Here are some calculations.
Without using your calculator, decide which are incorrect.

a $345 \times 0.2 = 69$

b $654 \div 3.7 = 1767.5675$

c $0.9 \times 0.09 = 0.81$

d $96 \times 54 = 5184$

e $531 \times 7211 = 3\,821\,300$.

Check with your calculator and correct the wrong ones.

UNIT 7 Number calculations

A Numbers without a calculator

While it is sensible to use a calculator when necessary, it is *important* to be able to do simple calculations in your head. It is often *quicker* to do them in your head.

Using pen and paper but NO calculator, complete Quiz 1. Time yourself.

Then complete Quiz 2.
Again, time yourself.

Quiz 1

1 Add together 57, 63 and 19.

2 Multiply 12 by 11.

3 Subtract 37 from 120.

4 Divide 51 by 3.

5 Find $12\frac{1}{2}\%$ of 200.

6 Add together $\frac{1}{2}$ and $\frac{1}{8}$.

7 Find $\frac{3}{4}$ of 1600.

8 How many minutes are there in $1\frac{1}{4}$ hours?

9 What is the cost of 24 pencils at 6 p each?

10 How long will it take to travel 75 miles at 30 miles per hour?

Quiz 2

1 Add together 52, 68 and 17.

2 Multiply 9 by 13.

3 Subtract 43 from 130.

4 Divide 91 by 7.

5 Find 15% of 300.

6 Add together $\frac{1}{3}$ and $\frac{1}{6}$.

7 Find $\frac{1}{5}$ of 1200.

8 How many seconds are there in 8 minutes?

9 What is the cost of 18 pencils at 5 p each?

10 How long will it take to travel 125 miles at 50 miles per hour?

Check your answers. Keep a record of your scores and times.

Repeat the quizzes in about a week's time. See if your scores improve.

Make up a quiz like the ones above. Exchange quizzes with a partner and see how well you can do your partner's questions.

B School timetables

Here is a pupil's timetable:

Lessons	1	2	3	4	5	6
Monday	Maths	English	French	History	PE	Geography
Tuesday	RE	Art	Art	Music	Maths	English
Wednesday	Home Ec	Home Ec	RE	History	Games	Games
Thursday	English	French	Maths	Maths	Science	Science
Friday	Science	Science	Geography	English	Technology	Technology

All lessons are 50 minutes long.

B1 How much time is spent on Maths each week?
What is the total time spent on lessons each week? What percentage of the lesson time for the whole week is taken by Maths?

B2 What is the ratio of the time spent on Music to the time spent on History and Geography together?

B3 How many minutes does this pupil spend on lessons in a week? How much time would he or she spend in a year of 40 weeks at school?

B4 How much time would the pupil spend on each subject if he or she had a similar timetable over a period of five years (of 40 weeks each year)?

B5 If the school changed from a 40-week year to a 38-week year, calculate:

a the percentage decrease in lesson time

b the percentage decrease in the total time for Maths.

B6 The school changes the timetable so that lessons 1, 2, 3 and 4 are now 55 minutes long and lessons 5 and 6 are 45 minutes long.

a Which subjects have an increase in time?

b Which subjects have a decrease in time?

c Calculate the percentage increase/decrease in time for each subject.

B7 Now use your own school timetable.

a What is the total time spent at lessons in a week?

b Make a table showing the time spent on each subject.

c Calculate the percentage of the total time spent on each particular subject.

d Represent the time spent on each subject on a chart (such as a pie chart).

e Calculate the time you spend on each subject in a 40-week year.

f Calculate the time you would spend on each subject if you had the same timetable for five years.

C Journey timetables

Here is part of a railway timetable:

London King's Cross	1030	1130	1235	1325	1430
Peterborough	1124	1235	1327	1415	1545
Newcastle	1336	1458	1539	1705	1730

C1 How long does it take the 1030 from London to reach Peterborough?

C2 What is the difference in the time between the fastest and the slowest trains from London to Newcastle?

C3 A traveller goes three times a week from Peterborough to Newcastle. What is the time spent on this train journey in a week:

a if the quickest possible train is used

b if the slowest possible train is used?

(Assume all trains run on time. Ignore the return journeys.)

Repeat **a** and **b** for a year of 48 weeks of travelling.

C4 Because of engineering work, all trains from London to Newcastle have their journey times increased by 10%. Work out the new arrival times in Newcastle for the trains in the timetable.

C5 A passenger, whose watch loses $1\frac{1}{2}$ minutes every hour, gets on the 1235 train at King's Cross and sets his watch to the correct time. To the nearest minute, what time will the watch show at Peterborough and at Newcastle, assuming the train runs on time?

D Money

D1 Make a list of the coins we use.

What is the smallest number of coins you could use to make a total of:

a 75p **b** £1.30 **c** £5.15?

What is the largest number of coins you could use to make a total of:

d 5p **e** £1.05 **f** £5.89?

Before 1971, in the United Kingdom, people used a system of money based on pounds (£), shillings (s) and pence (d):

£1 = 20s
1s = 12d
and so £1 = 240d

Sums of money were written in the form:

£12 5s 6d (twelve pounds, five shillings and six pence)
or 3s 4d (three shillings and four pence)

Some amounts of money had special names. For example:

A **guinea** was equal to £1 1s 0d
A **half crown** was equal to 2s 6d
A **florin** was equal to 2s 0d.

Electronic calculators were not widely available until about 1975, but even if they had been, they would not have helped you very much with this currency.

At the time of the change to our present decimal currency in 1971, the pound kept its same value and so:

- 100p (new pence) = 240d (old pence)
- and thus 1p = 2.4d
- and 5p = 1s

To convert 'old' money to decimal currency, the pound part stays the same. You need to look at the shillings and pence. For example:

Convert 4s 6d to pence.

1 shilling is the same as 5p, so 4s becomes $4 \times 5\text{p} = 20\text{p}$;
6d (half a shilling) is $2\frac{1}{2}$ p, making $22\frac{1}{2}$ p altogether.
(Since the $\frac{1}{2}$ p disappeared in the 1980s, there is no exact decimal equivalent today.)

D2 Convert these amounts of 'old' money to our decimal system:

a 2s **b** £1 10s **c** £2 3s 6d **d** £12 17s 6d.

D3 Convert these amounts of 'new' money to £.s.d:

a 50p **b** £2.60 **c** 72p **d** £3.72.

D4 **a** What are the values of the guinea, the half crown and the florin in our present system?

b You may have heard of horse races called the '2000 guineas' and the '1000 guineas'. What are the values of 2000 and 1000 guineas in decimal currency?

Extension work

D5 Add together £5 6s 3d and £6 15s 9d.

D6 Find the cost of 56 pencils at $3\frac{1}{2}$ d each. Give your answer:

a in £.s.d. **b** in 'new' pence.

The following questions appeared on an Arithmetic test paper for 10- to 11-year-old pupils in 1937. See how well you can do these strange questions!

D7 A man earns £363 13s 6d in one year. How much is this for each week?

D8 A shopkeeper's takings for a fortnight were £268 17s 6d.
During the first three days he took £39 19s 11d, and in the next three days he took £63 17s 8d. (Six days make one working week.)
How much more did he take in the second week than in the first week?

D9 I bought 97 $1\frac{1}{2}$d stamps together with some $2\frac{1}{2}$d stamps at the Post Office, spending in all £1 5s $0\frac{1}{2}$d. How many $2\frac{1}{2}$d stamps did I buy?

The next questions were originally published in a book dated 1886. See if you can do them!

D10 How many £1 17s 6d are there in £28 2s 6d ?

D11 A man's wages for 28 days are £4 0s 6d. What does he earn per day?

D12 A postman whose pay is 15s for a week is fined 1s 6d if he comes in late, and at the end of 13 weeks he receives £8 15s 6d.
How many times was he late?

E More problems on time and distance

Extension work

E1 How long would it take you to count to 1 million if you took 1 second to say each number?

If you started now, when would you finish?

If you were just finishing, when would you have started?

E2 **a** Find out how many times your heart beats in one minute:

i when you are sitting
ii after exercise.

b If you assume that your average heartbeat is the mean of the two rates that you obtained in **a**, how many times will your heart beat if you live to be 85 years old?

E3 **a** If people sleep for 8 hours a day on average, how many hours will they sleep in a lifetime of 80 years?

b Write this time in:

i seconds
ii minutes
iii days
iv weeks.

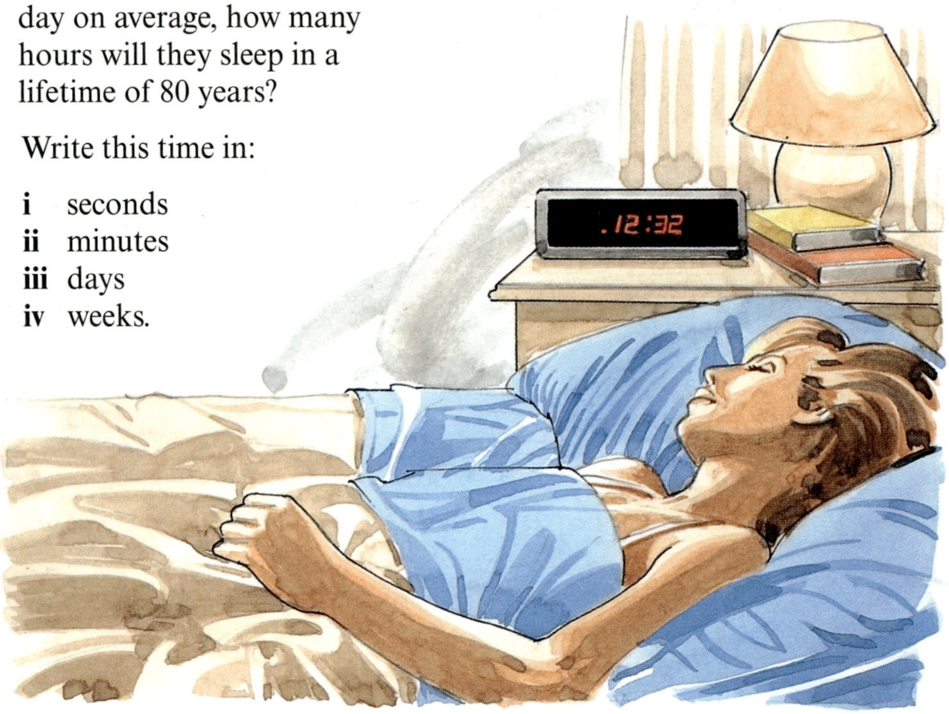

E4 The coronation of Elizabeth II was on 2 June 1953 at about midday. How many hours passed between then and midday on 1 January 1994? (Remember leap years!)

E5 For what percentage of the 20th century have you lived?

E6 A prisoner is given a choice: 'You can go free in either one year's time or when 10 million seconds have gone.'

Which should the prisoner choose?
(Work out your answer by two methods, and check that they agree.)

E7 If you placed a million pennies side by side, how far would they stretch? Estimate the answer first, then work it out.

How many pennies would be needed to make a mile for charity?

E8 Find the thickness of a page of a book by working out how many pages measure 1 cm.
How many sheets would you need to make a pile 1 km high?

E9 A jogger claims to run 5% further every day. If, on the first day, the jogger goes 1 mile, what will be the distance covered on:

- **a** the fifth day
- **b** the tenth day
- **c** the 50th day?

E10 Select a book (for example, a novel). Estimate the number of words in the book.

Now estimate the number of letters.
If a slow typist were to type these letters at a rate of one every second, how long would it take to finish the book?

UNIT 8 Coordinates and equations

A Extending coordinates

We can extend coordinates by using negative numbers.

Here is one quarter of a design:

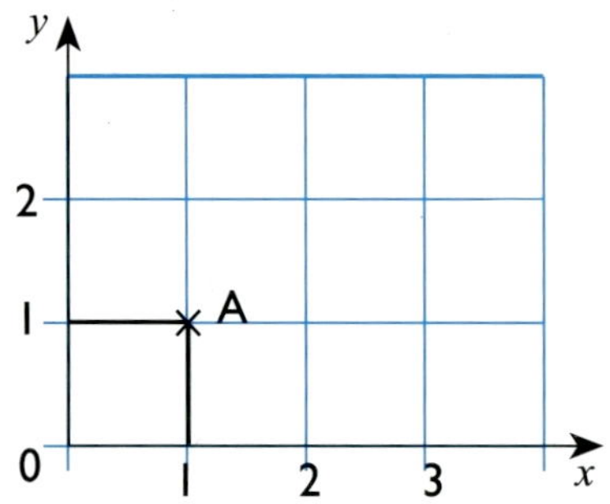

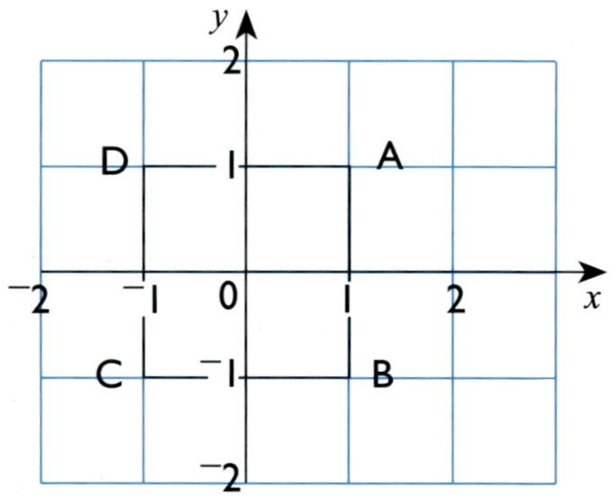

A1 Find the coordinates of the points B,C and D using the scales marked. List them in the form B (1,–1), etc.

Note carefully how we use negative numbers on the axes.

centimetre squared paper

A2 Here is one quarter of another design. Copy the design on to centimetre squared paper and then complete it. Find the coordinates of the points for the three new parts. Label all points carefully (with letters) and write down their coordinates alongside the letters.

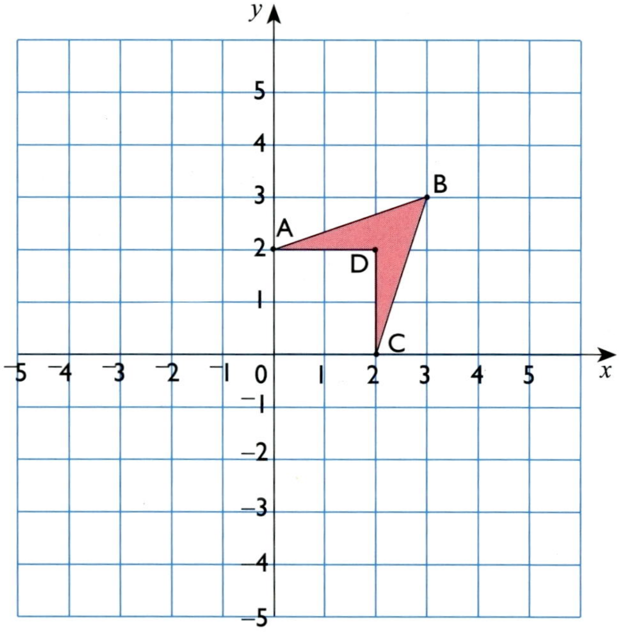

A3 Write down, in a table like the one below, the coordinates of the lettered points in the diagram.

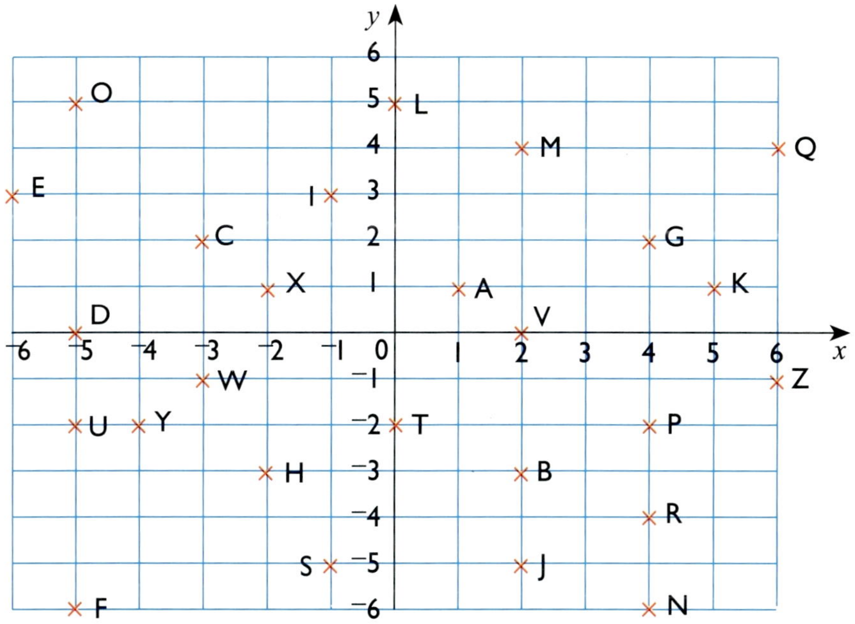

Column 1 (Positive X, positive Y)	Column 2 (Positive X, negative Y)	Column 3 (Negative X, positive Y)	Column 4 (Negative X, negative Y)
A (1,1)	J (2,–5)		

What do you notice about points in the same column?
Which quarters of the diagram do they come from?
We call these quarters **quadrants**.

A4 Complete the diagram below when it is reflected in the *x*-axis.

Label all the new points and write down their coordinates.

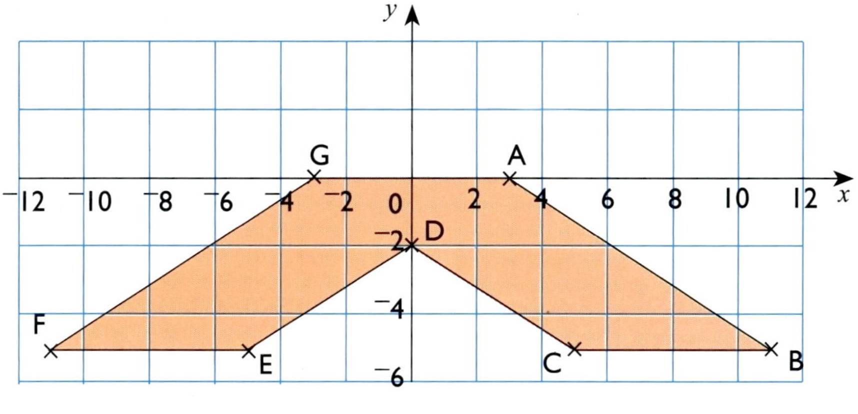

A5 Draw an x-axis with a scale from –6 to 12 and a y-axis with a scale from –8 to 8.

Plot these points and join them up in the order in which they were plotted:

A (6,7) B (5,1) C (11,–1) D (5,–2) E (4,–8)
F (1,–3) G (–4,–4) H (0,0) I (–3,5) J (2,3).

Join B, D, F, H, J to the point (3,0).

A6 You will need Worksheet N3:4.
Use five grids like this one.
Plot the five sets of points below, one on to each grid.

Join up the points as you plot them.

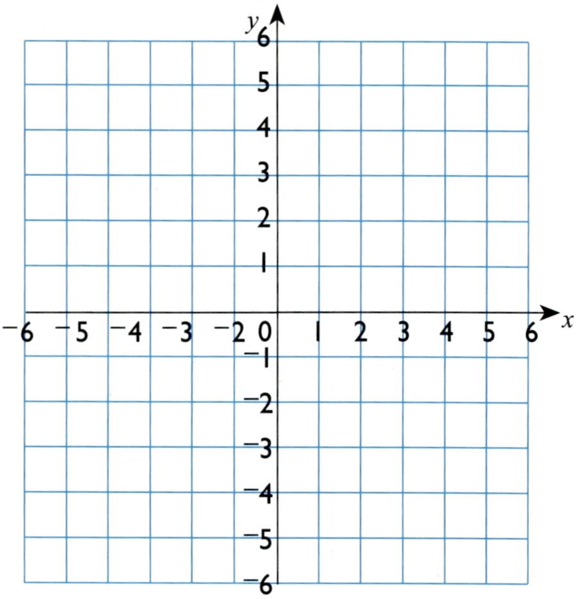

Name each shape.

a (4,–3) (2,2) (–5,1) (4,–3)

b (–4,–4) (4,2) (4,4) (2,4) (–4,–4)

c (0,5) (–5,0) (0,–5) (5,0) (0,5)

d (4,2) (2,4) (–2,4) (–4,2) (–4,–2) (–2,–4) (2,–4) (4,–2), (4,2)

e (1,–5) (4,–2) (–2,4) (–5,1) (1,–5)

A7 Using centimetre squared paper draw x- and y-axes. Make a scale from –3 to 5 on each axis using 1 cm for 1 unit.
Plot the points A (1,2), B (1,5) and C (–3,2).
Join them. What type of triangle have you produced?

A8 Draw an x-axis from –10 to 14 and a y-axis from –6 to 10. Plot these points (they all lie on a straight line):

(–9,–6) (–5,–2) (0,3) (2,?) (4,?) (?,8) (?,10)

B Using the computer

LOGO

> The following work uses LOGO. Remember that:
>
> **setpos** means **set the position**.
>
> Coordinates are placed inside square brackets.
>
> Coordinates are calculated from the centre of the screen.
>
> **pu** means **pen up** which allows you to move around the screen without leaving marks.
>
> **pd** puts the **pen down** again.

B1 Load your LOGO program, then type this:

```
pu
setpos [-150 0]
pd
setpos [0 150]
setpos [150 0]
setpos [0 -150]
setpos [-150 0]
```

Now draw another square around this one. Start with:

```
pu
setpos [-250 0]
pd
setpos [0 250]
```
... and so on.

B2 Using LOGO, draw two squares like this. Record your LOGO instructions carefully.

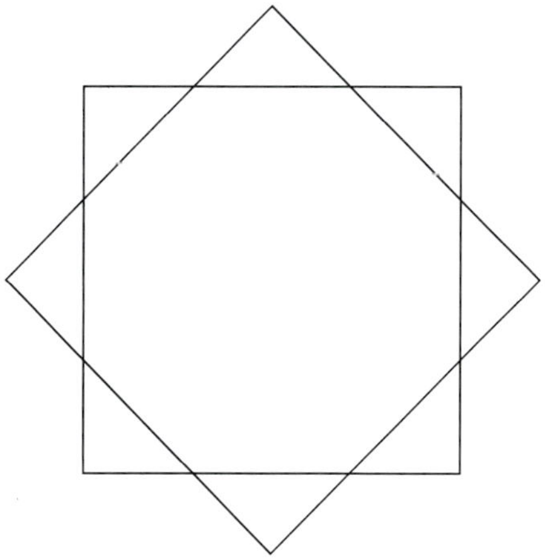

Here are some BASIC programs. Before you run them write down what you think they will do.

B3

```
10 MODE 1
20 INPUT X
30 INPUT Y
40 DRAW X,Y
50 GO TO 20
```

What does the program do? Try positive and negative values for X and Y.

B4

```
10 MODE 1
20 INPUT X
30 LET Y = X
40 DRAW X, Y
50 GO TO 20
```

Run this for 10 values of X. Explain what is happening.

B5

```
10 MODE 1
20 LET X = 0
30 LET Y = X
40 DRAW X,Y
50 LET X = X + 1
60 GO TO 30
```

What does this do?

B6

```
10 MODE 1
20 DRAW 0,0
30 DRAW 0,40
40 DRAW 60,0
50 DRAW 60,40
```

What shape do you get?

B7

```
10 MODE 1
20 MOVE 200,300
30 DRAW 900,300
40 DRAW 900,700
50 DRAW 200,700
60 DRAW 200,300
```

What shape do you get?

B8 Find the effect of this program:

```
10 MODE 1
20 FOR Y=0 TO 600 STEP 100
30 DRAW 0,0
40 DRAW 0,Y
50 DRAW 100,Y
60 DRAW 100,0
70 NEXT Y
```

Run the program. Now extend the program by adding:

```
80 FOR X=0 TO 600 STEP 100
90 DRAW 0,0
100 DRAW X,0
110 DRAW X,100
120 DRAW 0,100
130 NEXT X
```

Now write your own program to produce a set of triangles.

C Linear equations

Use a set of x- and y- axes with values of x and y ranging from –5 to +5 in cm units to answer the following questions (a fresh grid for each question is not always necessary):

C1 Draw a straight line to join the points (4,5) and (–2,–3). How long is the line? What are the coordinates of its mid-point?

C2 Mark on a grid four points which are 2 cm from the y- axis and 4 cm from the x- axis. Give the coordinates of the points.

C3 Make a list of at least six points which are equal distances from the x- and y- axes. Mark them on a grid.

C4 If the point (3,0) is one corner of a square, give the coordinates of three other points which could be corners of the same square. Draw the square.

C5 Mark on a grid at least four points which are twice as far from the y- axis as from the x- axis. Give the coordinates of the points.

C6 Shade the area where all the points which are 1 cm or less from the origin (0,0) will be.

C7 (–3, –3) and (0, 0) are opposite corners of a square. What are the coordinates of the other two corners? Draw the square.

C8 The coordinates of the point (3, 1) add up to 4. Find six more points that have coordinates adding up to 4. Mark them on a grid. (Use some points which have fractional coordinates and some points which have negative coordinates.)

C9 Shade in the area where the points have x- coordinates greater than y- coordinates.

C10 Shade in the area where points have y-coordinates less than 5.

C11

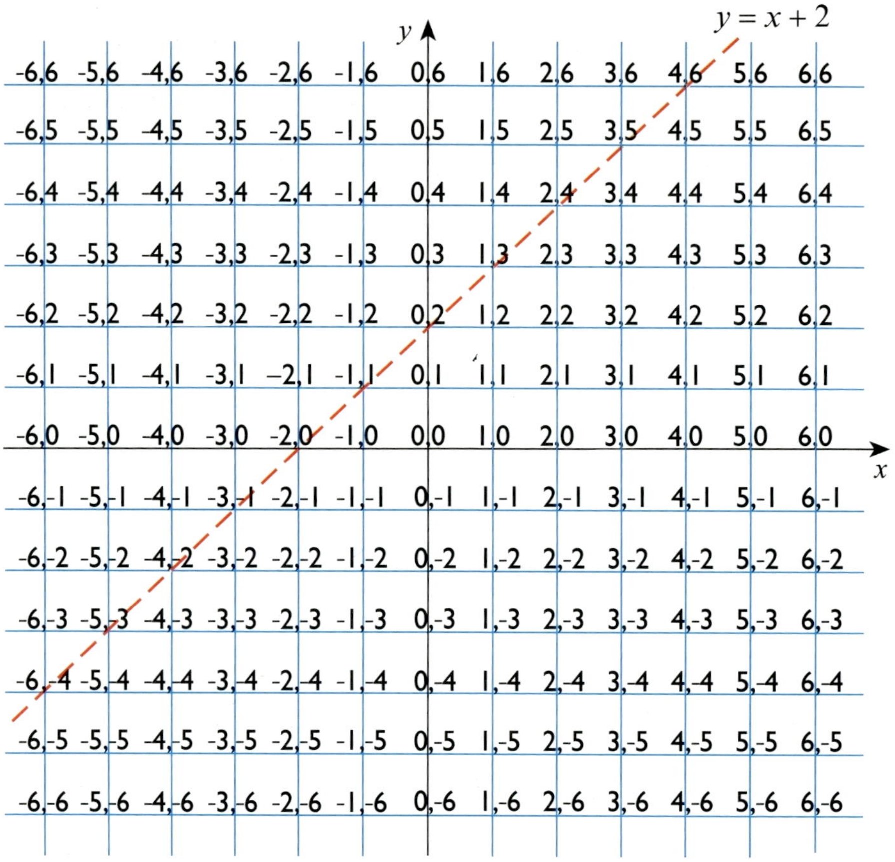

Here is a coordinate grid with whole number coordinate points marked. A dotted line has been drawn through the points (4, 6), (3,5), (2,4), (1,3) … Notice that for all these points the ***y*-coordinate is 2 more than the *x*-coordinate**. Because of this, **the equation of this line is $y = x + 2$.**

Check that all the other points which the line passes through have y-coordinates which are 2 more than the x-coordinates.

C12 List all the marked points on this grid with y-coordinates which are 4 more than the x-coordinates. What is the equation of the line that passes through them?

C13 List the points marked on the grid and the equations of the lines passing through them if:

a the y-coordinates are 4 less than the x-coordinates

b the y-coordinates are equal to minus the x-coordinates

c the x- and y-coordinates add up to 3

d the x- and y-coordinates add up to –4

e the y-coordinate is twice as big as the x-coordinate

f the y-coordinate is half as big as the x-coordinate.

C14 Place a piece of tracing paper over the diagram. Mark the x- and y- axes. Mark in the lines for the equations:

a $y = x$ **b** $y = x–2$ **c** $y = 3x$ **d** $y = –x$ **e** $y = –2x$.

C15 If a straight line passes through the point (3, 6) make two suggestions for its equation.

C16 If a straight line passes through (3, 1) and (–3, –1), name two other points it passes through. What is the equation of the line?

C17 Name a point on the grid that the line $y = 2x + 1$ will pass through.

C18 The line for the equation $y = 5$ passes through all the points which have a y-coordinate of 5 (the x-coordinate can be anything). What is the position and direction of the line in relation to the x- and y-axes?

C19 Describe the position and direction of the line for ($x = 3$) in relation to the x- and y-axes.

C20 Sketch coordinate grid axes. On the sketch:

a draw the line for $y = 3$

b shade the area where $y > 3$

c shade (in a different colour) the area where $y < 3$.

C21 Sketch coordinate grid axes. On the sketch indicate the following regions using different colours or shadings:

a the area where $x > 4$

b the area where $y > 4$

c the area where both $x > 4$ and $y > 4$

d the area where both $x < 4$ and $y < 4$.

D Drawing graphs

When you want to draw a graph for an equation, it is not always possible to pick out the points from a grid that will fit the equation. This is because grids do not have all the coordinates of points showing and you may, for example, want to plot a point with coordinates like (1.5, 3.25).

It is nearly always best to take the equation, put values for x into it and work out the corresponding values for y. Then plot the points and join them up to make the line.

For example, to plot the graph of $y = 2x - 5$:

Choose a range of values for x, say from –3 to +3. Put each value into the equation and work out a value for y:

If $x = 3$, $y = 2 \times 3 - 5$

$= 1$

so (3, 1) is a point.

If $x = 2$, $y = 2 \times 2 - 5$

$= -1$

so (2, –1) is a point.

If $x = 1$, $y = 2 \times 1 - 5$

$= -3$

so (1, –3) is a point.

And so on.

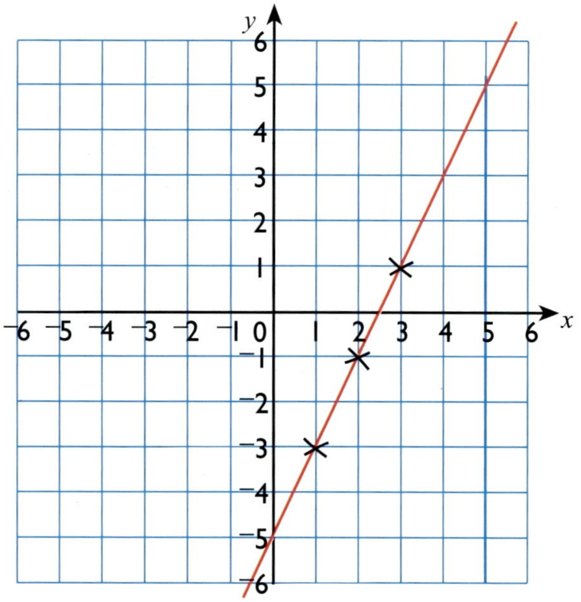

Some of the points may be off the grid. Once the line is fitted to the points it can be extended in either direction. We normally find *three* points to draw a straight line, two points will fix the line and the other gives a check that no error has been made.

D1 Copy the axes and points in the diagram above. Label the line $y = 2x - 5$ and write down the coordinates of three other points on the line.

D2 For the same range of values of x (from –3 to +3), and with the same axes, work out three points for each of the following equations and draw the graph of each one:

(in each case choose one of your points with $x = 0$)

a $y = 3x + 1$ **b** $y = \frac{1}{2}x + 3$ **c** $y = -x + 1$.

D3 On a new grid, over the same range of x, plot graphs for:

a $y = \frac{1}{2}x - 5$ **b** $y = 1 - 2x$ **c** $y = 4x - 1$.

D4 On a new grid, plot the graph of $y = \frac{1}{2}x - \frac{1}{2}$ and the graph of $y = 4x - 4$. What are the coordinates of the point where they cross each other?

D5 Find the coordinates of the crossing point (often called 'the point of intersection') of the graphs:

$y = 4 - x$ and $y = 4 + x$.

If you have a graphical calculator available, use it to help you with the next three questions.

D6 Using a graphical calculator, or by hand on the same axes, plot the following graphs over a range of x from –5 to +5. Label them clearly.

a $y = 2x + 5$ **b** $y = \frac{1}{2}x - 2$ **c** $y = 2x - 1$ **d** $y = \frac{1}{2}x + 3$.

D7 Examine the graphs in question **D6** and say what:

a the graphs of $y = 2x + 5$ and $y = 2x - 1$ have in common

b the equations of $y = 2x + 5$ and $y = 2x - 1$ have in common

c the graphs of $y = \frac{1}{2}x - 2$ and $y = \frac{1}{2}x + 3$ have in common

d the equations of $y = \frac{1}{2}x - 2$ and $y = \frac{1}{2}x + 3$ have in common.

D8 Make up your own equations and find ones that are:

a parallel to $y = 2x + 5$

b parallel to $y = \frac{1}{2}x + 3$.

UNIT 9 Linear equations

A Problems

A1 **a** Three oranges and one banana cost 50p. The banana costs 14p. How much is each orange?

b I paid £1 for six bread rolls and a cake and got 28p change. The cake cost 24p. How much were the bread rolls?

c Ben owed Mabel 25 marbles. She accepted seven marbles and nine toffees from him. How many marbles is a toffee worth?

d A motorist bought 6 gallons of petrol and 5 litres of oil for £20.30. The oil cost £6.50. How much was petrol per gallon?

e Two colas and a 75p packet of popcorn cost £2.05. How much does one cola cost?

f Three people shared two pizzas in a restaurant. The total bill was £6.60. They each put in the same amount of money but found they were 48p short of the total bill. How much had each person contributed?

g Eight carpet squares will just fit along the width of a floor if a 7cm strip is cut off one of the squares. The width of the floor is 2.33 m. How wide is a carpet square?

h A sweet manufacturer makes chocolate Peanut Twizzles. A box of 15 is 50 g short of 1 kg. What is the mean weight of a Peanut Twizzle?

i Each post in this fence is 25 cm wide. How wide is the gap between each post?

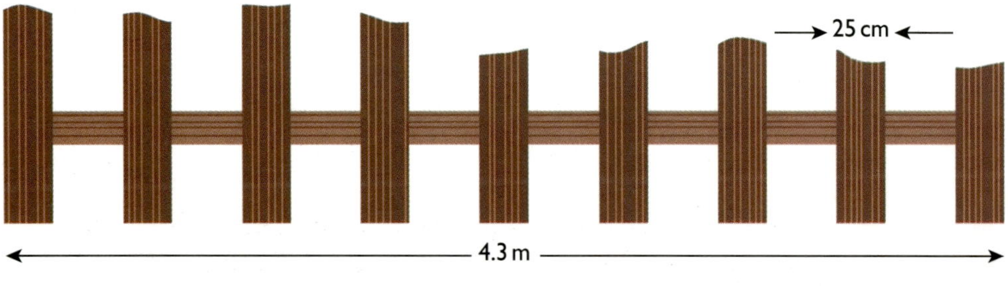

A2 Find the original number in each of the following:

a Twice a number is 93.

b Half a number is 24.

c Add 4 to a number and you get 96.

d Add 1 to a number and you get zero.

e Take 10 away from a number and you get 117.

f Five times a number is 15.

g Five times a number is 1.5.

h Five times a number is 0.15.

i Eight times a number is 272.

A3 In question **A2 a**, if you let 'x' be the original number, then the statement: 'twice a number is 93' can be represented by the equation: $2x = 93$

For each of the other statements in question **A2**, write an equation with 'x' in it for the original number.

A4 Put these equations into words using 'a number' in place of 'x' and write down what you think the number is in each case:

a $7x = 28$ **b** $x - 30 = 25$ **c** $3x + 8 = 23$ **d** $2x - 12 = 1$.

A5 Mehmet was asked to think of a number, add 5 to it and multiply the answer by 2. His answer was 43. What was his number?

A6 Flora thought of a number, subtracted 3 from it and then multiplied the answer by 4. She got the answer 100. What was her number?

A7 A confusing mathematician said his dog weighed half its own weight and 5 kg. What did his dog weigh?

A8 Lulu added 50p to half her savings so that she could buy a book for £7.25. How much had she saved?

A9 What was the original price of this television set?

A10 Find the number which, when you double it and add 1, gives the same answer as if you just added 8.

A11 Find the number which, when you halve it, gives the same answer as if you had simply subtracted 20 from it.

B Balancing and equations

Cherie asked her younger sister, Deli, to weigh a parcel for her, using a balance and some kilogram weights. Deli made the scales balance, as you can see in the diagram, but she couldn't work out the weight of the parcel.

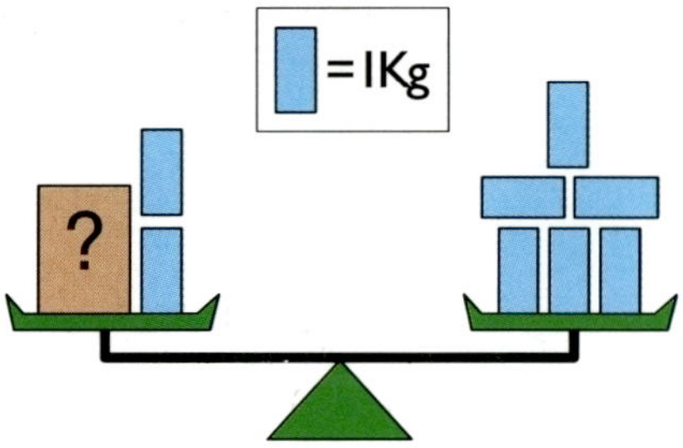

Cherie said: 'It's easy – you just take two kilogram weights off each side, so you've got the parcel on its own on one side, so you've got the parcel on its own on one side. Now you can see that the parcel must weigh 4 kg.'

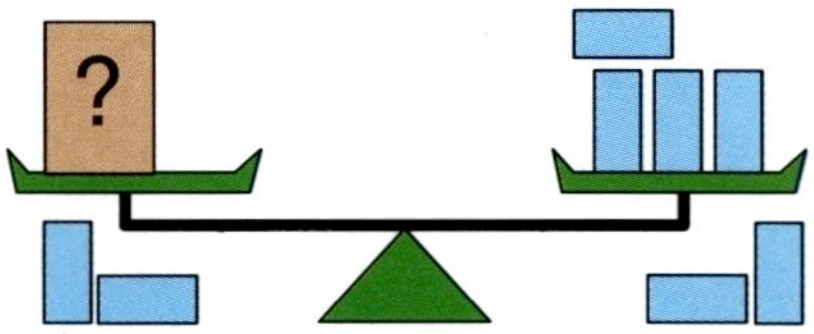

'OK' said Deli, 'how would you do this then?' as she balanced the scales with two identical parcels and some kg weights.

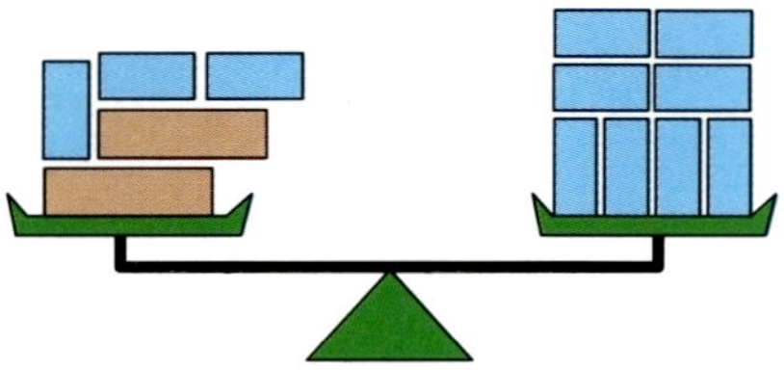

'That's no problem. You just take 3 kg off each side, so you know two parcels weigh 5 kg. Then divide by two to find the weight of one parcel – which must be 2.5 kg' said Cherie.

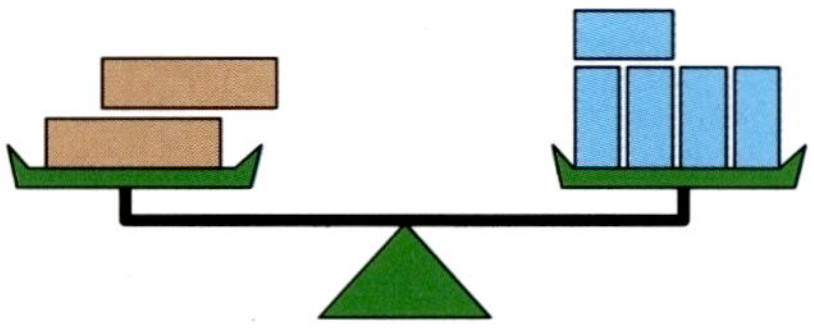

Instead of drawing a diagram, you could use a mathematical equation to represent the balanced scales.

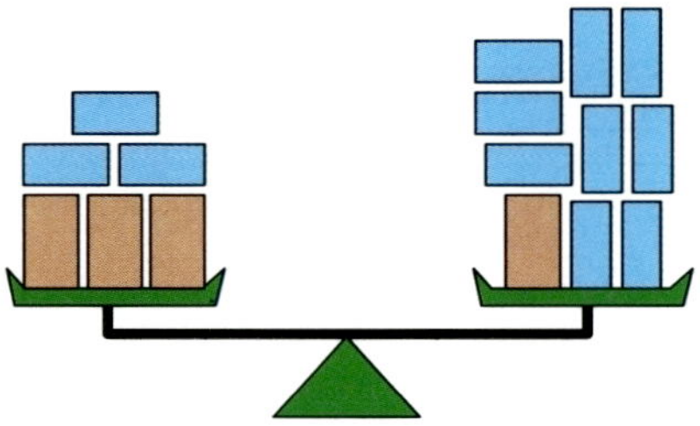

In this diagram, 3 parcels + 3 kg balances 1 parcel + 9 kg

or $3p + 3 = p + 9$ (if you let p stand for the number of kg a parcel weighs)

Take 3 kg off each side: $3p = p + 6$
Take a parcel off each side: $2p = 6$
Divide by 2: $p = 3$

In other words, the parcel weighs 3 kg.

B1 Write an equation for this balancing scale and write down the steps to solve it (and therefore find the weight of a parcel).

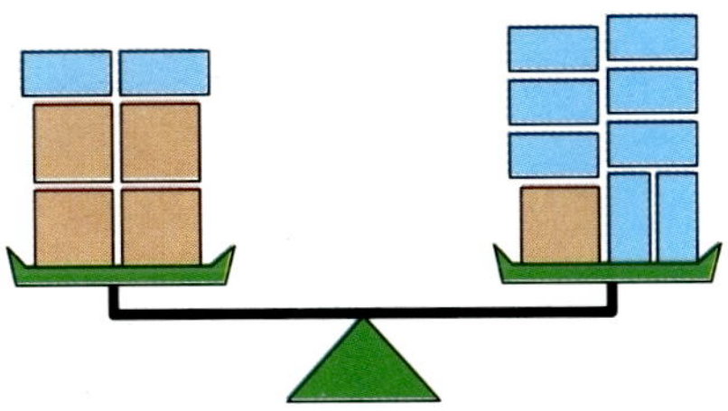

B2 Write an equation for each of the following and then solve each one:

a five parcels and 2 kg balance 12 kg
b three parcels and 5 kg balance one parcel and 12 kg
c ten parcels and 3 kg balance two parcels and 15 kg.

B3 Draw a sketch of balancing scales to fit the equation $3p + 1 = 2p + 7$. Solve the equation.

B4 Solve these equations:

a $3p + 5 = 20$ **b** $6p + 7 = 31$ **c** $4p + 1 = 2$

d $7p + 3 = 38$ **e** $9p + 1 = 7p + 5$ **f** $4p + 5 = p + 14$

g $8p = 28$ **h** $4p = 25$.

B5 You have three parcels weighing 2 kg each and three weights: 5 kg, 4 kg, and 7 kg.

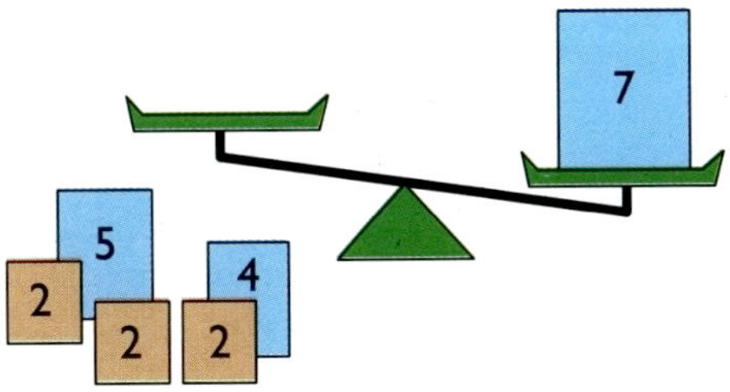

Find two different ways that they can be arranged on the scales so that they balance. Write down the equation for each arrangement.

(Use p to represent the number of kilograms in the weight of a parcel again.)

A cheese problem

A grocer had two identical blocks of cheese. During the day he cut 350 g off one of them to sell.

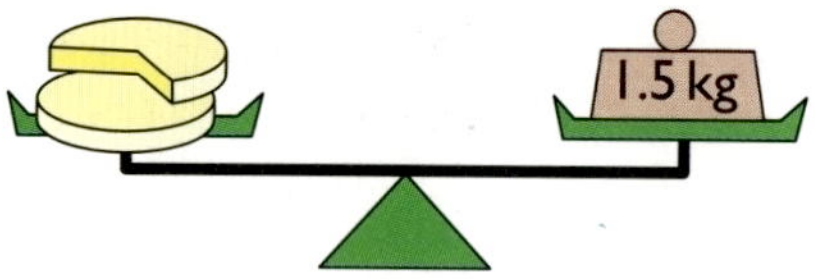

When he weighed the two blocks afterwards, their combined weight was 1500 g. How much did each block weigh at the start?

You could write an equation for this: $2c - 350 = 1500$ (where 'c' stands for the weight of a block of cheese in grams).

To solve this equation, imagine adding 350 g to each side of the scales. They will still balance since the same weight is added to both sides.

The equation is now:

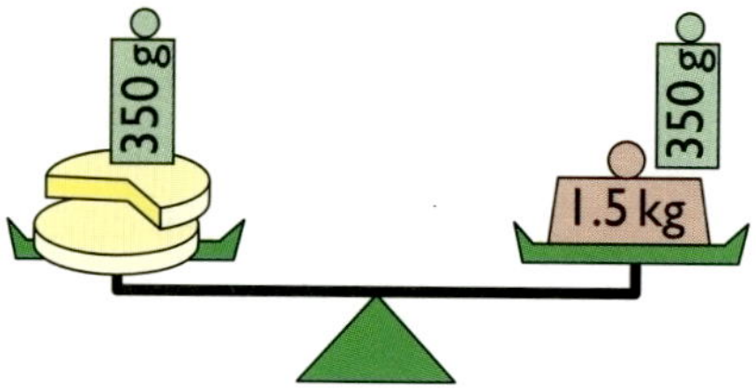

$2c - 350 + 350 = 1500 + 350$

which is: $2c = 1850$

so: $c = 925$.

Each block of cheese weighed 925 g.

B6 If the equation in the cheese problem had been $2c - 400 = 900$, how would you have solved it?

B7 Solve these equations:

a $5c - 200 = 750$ **b** $3c - 20 = 19$ **c** $4c - 5 = 27$ **d** $10c - 1 = 14$.

To solve an equation you need to rearrange it so that you have the unknown (such as x or c) on its own on one side of the equation. To do this you can add, subtract, multiply or divide but whatever you do *you must do the same to both sides of the equation*.

For example: $3x + 7 = 16$
Subtract 7 from both sides: $3x = 9$
Divide both sides by 3: $x = 3$

You do not always need all of these rules to solve an equation. It is usually best to apply these rules in the order they are given above.

B8 Solve these equations:

a $3h + 5 = 2h + 13$ **b** $4g - 7 = 2g + 11$ **c** $5q + 3 = 18$

d $2y - 10 = 23$ **e** $t + 13 = 4t + 1$ **f** $8w + 5 = 12w - 21$

g $10x = 1$ **h** $9p + 2 = 2p + 30$ **i** $x - 1 = 3x - 2$

j $8z - 5 = 10z - 8$.

B9 The radio-cassette is worth £67.50. Write an equation using 'b' for the value of a book of stikkistamps. Solve the equation to find out how much a book of stikkistamps is worth.

B10 In a factory store room, six packing cases laid side by side leave a gap of 14 m to the wall. Nine packing cases leave a gap of 9.5 m to the wall.

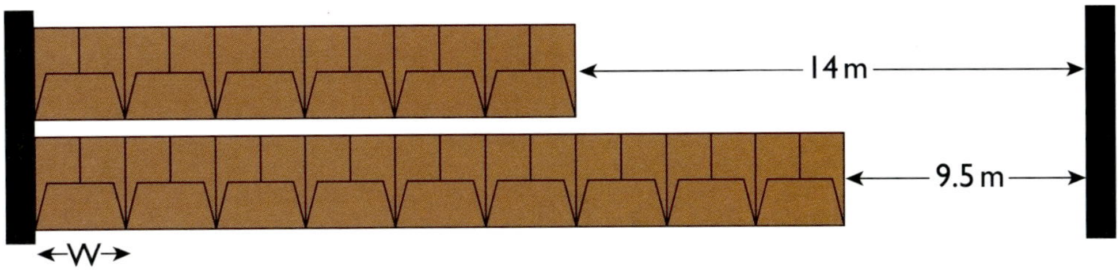

Call the width of a packing case w and make an equation out of this information. Solve the equation to find the width of a packing case.

B11 I am thinking of a number. When I double it and add 10, I get the same answer as multiplying it by 5 and subtracting 8. Write an equation for this and solve it to find the number.

C Another way to solve equations

Ruby asked Zaluf to: think of a number, double it and add 20. Zaluf said the answer was 46. 'Then your number must have been 13,' said Ruby. 'How did you know that?' asked Zaluf. 'I just worked it out backwards,' Ruby said, 'like this:'

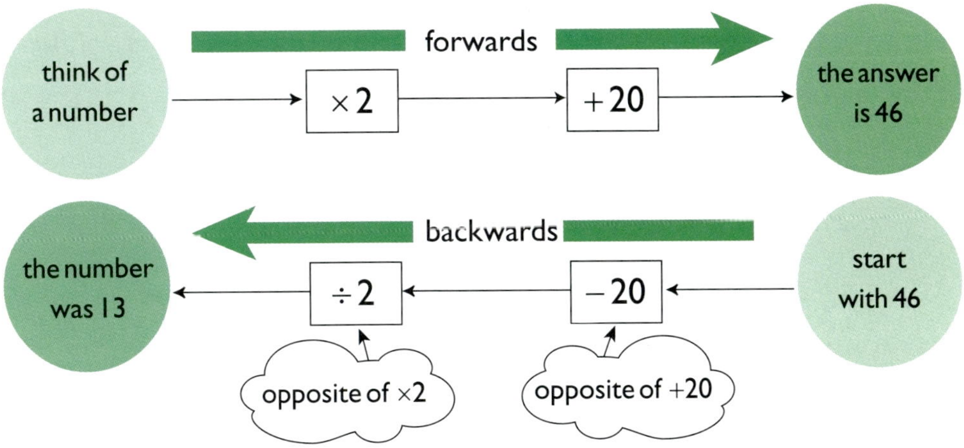

C1 Draw a diagram for: think of a number, multiply by 3, add 8 and the answer is 41.

C2 Draw the diagram for 'going backwards' in question **C1**, starting with 41, and work out what the original number must have been.

C3 In the following diagrams, 'n' stands for the original number. Draw the backwards diagrams and find each value of n.

a $(n) \to [\times 5] \to [+4] \to (39)$

b $(n) \to [\times 6] \to [-9] \to (51)$

c $(n) \to [\times 8] \to [+11] \to (15)$

d $(n) \to [\div 2] \to [+5] \to (91)$

e $(n) \to [\div 4] \to [+4] \to (8)$

f $(n) \to [+10] \to [\div 3] \to (17)$

You can use the same method to solve an equation such as:

$$3y + 5 = 38$$

'y' is the starting number; it has been multiplied by 3 and then 5 has been added.

$3y + 5 = 38$ is the same as $(y) \to [\times 3] \to [+5] \to (38)$

The backwards procedure is '-5' followed by' $\div 3$':

$$3y + 5 = 38$$

-5 from both sides: $3y + 5 - 5 = 38 - 5$

$$3y = 33$$

$\div 3$ both sides: $3y \div 3 = 33 \div 3$

$$y = 11$$

C4 Solve the following equations by the 'backwards method':

a $7w + 50 = 99$ **b** $2y - 44 = 65$ **c** $12f + 1 = 7$

d $4x - 1.7 = 2.9$ **e** $0.1y - 3 = 5$ **f** $1.4q = 7 = 35$

g $(p + 5) \times 7 = 42$ **h** $(r - 1) \times 3 = 27$ **i** $(w + 4) \div 5 = 15.$

D Mixed problems

D1 A rubber band is stretched so that it makes a rectangular shape, twice as long as it is wide. The total length of the stretched band is 27 cm. How wide is the rectangle?

D2 I bought three computer games at different prices. One was quite cheap. One cost twice as much as the cheap one, and the third cost three times as much as the cheap one. The bill for them all was £75. How much did they each cost?

D3

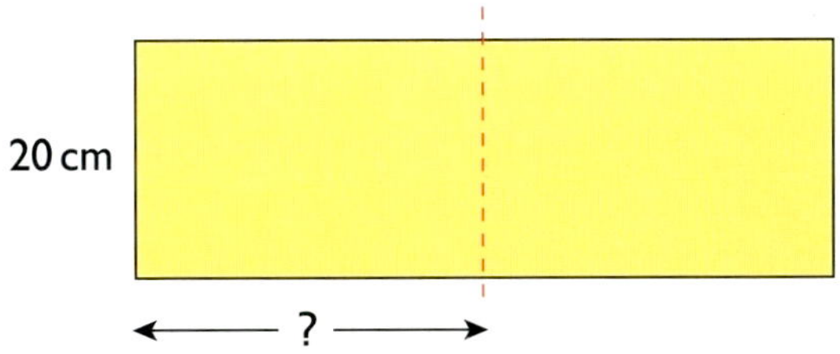

A piece of wood is 20 cm wide. If it has to be cut so that its perimeter is 108 cm, how long will it be?

D4 Three numbers add up to 100. The second number is twice as big as the first and the third number is 12 more than the first. Find the numbers.

D5 A boy had a ruler with a few centimetres broken off the beginning. He did not notice this when he measured the sides of a triangle as 12 cm, 15 cm and 17 cm. The perimeter of the triangle was really 36.5 cm. How much was broken off his ruler?

D6 The air freight charge on a flight to West Africa is £13.50 per kg plus a handling charge of £24. How many whole kg can you send for a maximum of £100?

D7 Solve the equations:

a $\frac{1}{2}x - 17 = 53$ **b** $\frac{1}{4}x + 5 = 6.1$.

D8 Two years ago a woman was twice as old as her daughter. The age difference between them is 20 years. How old are they now?

D9 A new size of bottle of bathroom cleaner contains 150 ml more than the old size. Three of the new bottles contain the same as five of the old ones. What was the volume of the old bottle?

D10 The three angles of a triangle are in the proportion 2: 3: 4. Find the angles.

D11 X stands for an odd number.

a Write down in terms of X the next two odd numbers after X.

b X and the two odd numbers after it make a total of 75. Find X.

D12 Membership of a film club costs £15 per year. Members pay £3.50 to see a film. Non-members pay £4.25 to see a film.

a Write expressions in terms of 'n', the number of films seen, for the cost of seeing films as a member and as a non-member.

b Last year I joined the club and saw enough films to make a saving of £9 (allowing for the cost of membership) compared with the cost if I had not been a member. How many films did I see?

D13

$5x - 2$	$\frac{1}{2}x$	$3x$
$x + 1$	$2x + 1$	$3x + 1$
$2x$	$6x - 3$	x

This is a magic square. All the rows, columns and diagonals add up to the same total.

Find the value of x to make this so, and from this, find the total of any row, column or diagonal.

D14 The formula $5n - 7$ is used to generate a sequence by putting in values of $n = 1, 2, 3 \ldots$

a What is the 50th term of the sequence?

b What number term of the sequence is 138?

c The formula $3n + 17$ will also generate a sequence What value of n makes each formula produce the same number?

D15 A box of Calibre Crescent chocolates has chocolates in it with three kinds of filling: nut, toffee and creme. In the standard box, there are three more toffee fillings than nut and twice as many creme fillings as toffee. If 'n' is the number of nut chocolates:

a write down expressions for the number of toffee and creme chocolates in terms of n

b find the value of n if there are 33 chocolates in the box.

E Practical examples of linear graphs

E1 A spice shop sells its own mix of curry powder for 80p per 100 g. Make a table to show the cost of 100, 200 … up to 600 g of curry powder. Plot a graph of weight (x-axis) against cost (y-axis). Use the graph to find:

a the cost of 120 g and 280 g of curry powder

b how much curry powder you can get for £2 and for £3.50.

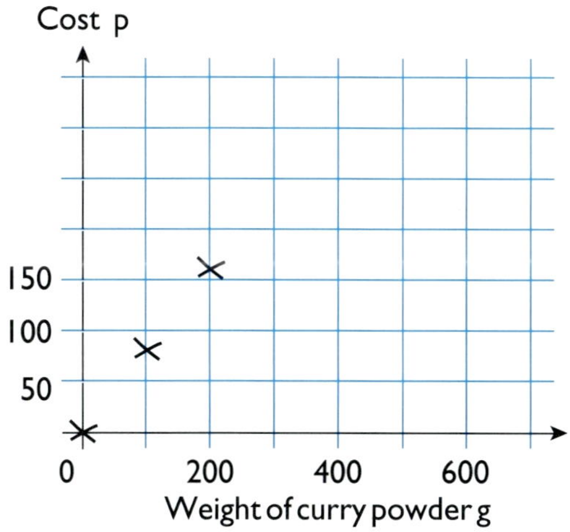

E2 A car travels an average of 16 km per litre of petrol. Another car with a specially adjusted engine can travel an extra 3 km per litre.

a Draw two graphs (on the same axes) to show how far the two cars can travel using up to 50 litres of petrol.

b Use the graph to find out how much further the second car can go with 36 litres of petrol.

c How much petrol will each car use to travel 500 km?

d Each car is filled with the same amount of petrol for a test run. At the end of the run the first car is 120 km behind the second. How far has each travelled?

E3 A pet-shop owner sells dried cat food in 15 kg sacks for £22. She also sells smaller quantities of the cat food loose at the same scale of pricing.

a Draw a graph of weight of cat food against cost which could be used to work out the cost of various amounts of cat food up to 15 kg.

b Use the graph to find the cost (to the nearest 10p) of 1 lb (2.2 kg) and 3.5 kg.

c If the shop owner decides to make up bags of cat food costing 75p and £2.50, what weight will be in each (to the nearest 10 g)?

UNIT 10 Mappings, functions and graphs

A The function game

This game is best played between one person and a group – such as the teacher (or a pupil) and the rest of the class. Draw two vertical lines on the blackboard to separate it into three parts. The left-hand and middle columns are for writing numbers and the right-hand side is for comments.

The teacher asks someone in the class to write a number (not too large – less than 20, say) in the left-hand column with white chalk. The teacher (with a particular function in mind – for example double the number and subtract 3), writes a response in the middle column in a different colour (yellow, say). From then on there is absolute silence.

Another person from the class writes a number on the left and the teacher gives his or her response in the middle column. This continues for a while, with the teacher offering the yellow chalk to anyone who wants to write in the middle column as well as the left-hand one. Wrong numbers are erased.

At any time in the game people are allowed to write comments on the blackboard about what they have noticed or what they think is the set of operations that the teacher is using to make the response each time.

Here is a game that has been played:

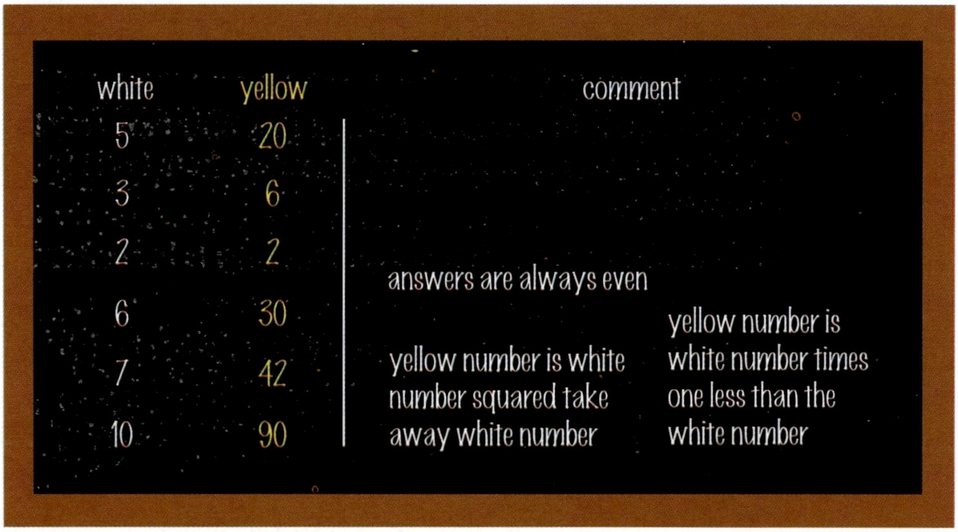

A1 Play the function game in your class a few times, either on the board or with a neighbour. Look at the comments that are made in your own games and see if you can decide which of them are correct. When there are two or more descriptions of the process for changing white numbers to yellow ones, see whether the ways are equivalent.

A2 Look at the game on the previous page and explain:

a why the answers are always even

b why $(\text{white})^2 - \text{white} = \text{white} \times (\text{white} - 1)$.

A3 Can you spot the rules used in these two games?

a

3	3
8	−2
1	5
10	−4

b

5	0
4	2
3	4
2	6

A4 When a class played the function game a few times, the following comments came up. Try to give an explanation for each. (They only apply to whole numbers.)

a When you double a number, the answer is always even.

b When you multiply a number by the number which is one more than it, you get an even number.

c If you add 5 to a number and then multiply by 2, you get the same answer as multiplying the original number by 2 then adding 10.

d Doubling a number then squaring it is the same as squaring the number then multiplying by 4.

e Taking a number then adding it to the one before it and the one after it is the same as multiplying it by 3.

f Halving a number then squaring it is the same as squaring a number then dividing by 4.

A5 On one occasion playing the function game, there were only two numbers on the blackboard – a white number and its matching yellow one. One person wrote, 'Yellow is twice the white number, add seven.' Another wrote 'Yellow is four times the white number take away 11.' Both explanations fitted the numbers. What were the numbers on the blackboard?

A6 Think of a number. Add 5. Double your answer. Add 2. Add your first number. Divide by 3. Subtract 4. Your answer is the first number you thought of. How does this trick work?

B Function machines

You could play the function game with a computer that had been programmed with a list of functions to use. It could choose a function at random from its list and check whether your suggestions for numbers were correct. Let us call such a computer a **function machine**. Here is an example of a function machine:

input 4 → [×2 +5] → 13 output

If you put y into the machine, the output would be $2y + 5$ so another way to describe the operation of this machine is:

$$y \dashrightarrow 2y + 5$$

This is called a **mapping**. We say 'y is mapped onto $2y + 5$'. It is like making a map of an area of land. For every place you identify on the land, you mark a point on the map (on a smaller scale). In a similar way, for every number you can think of there is another number you can make out of it by multiplying by 2 and adding 5.

B1 Find the missing numbers in the following function machine diagrams and say what the mapping is for each one (in terms of 'x'):

a 4 → [×5 +2] → ? **b** 10 → [÷2 –5] → ?

c 6 → [×? +5] → 29 **d** 5 → [×6 –?] → 14.

B2 For each of these machines write three different functions or mappings:

a 25 → [?] → 45 **b** 17 → [?] → 16.

B3 These two machines are operating with the same function or mapping. What could it be?

2 → [?] → 2 3 → [?] → 6

Can you find how this machine works?

Here are two function machines together:

$\longrightarrow$ [$\times 2$ $+4$] $\longrightarrow$ [$\times 3$] $\longrightarrow$

If we put 5 in, the output from the first will be $5 \times 2 + 4 = 14$, and this is the input to the second. The final output will be $14 \times 3 = 42$.

If we put 'y' in we get $2y + 4$ from the first machine. This is then multiplied by 3 in the second. $(2y + 4) \times 3 = 6y + 12$. So the two machines are equivalent to the mapping y ---► $6y + 12$.

B4 **a** If the two machines above were interchanged, would the output be the same from an input of 5?

b What is the single mapping equivalent to the two machines in this new order?

B5 Find the final outputs for each of these pairs of function machines. Write down a single mapping for each pair.

a 12 $\longrightarrow$ [$\times 6$] $\longrightarrow$ [$\div 2$] $\longrightarrow$

b 9 $\longrightarrow$ [$\times 2$ -7] $\longrightarrow$ [$+2$] $\longrightarrow$

c 11 $\longrightarrow$ [$\times 5$] $\longrightarrow$ [$\times 2$ $+1$] $\longrightarrow$

d 11 $\longrightarrow$ [$\times 2$ $+1$] $\longrightarrow$ [$\times 5$] $\longrightarrow$

e 20 $\longrightarrow$ [-3] $\longrightarrow$ [$\times 10$ $+2$] $\longrightarrow$

f 20 $\longrightarrow$ [$\times 10$ $+2$] $\longrightarrow$ [-3] $\longrightarrow$

B6 The following pairs of function machines always produce an output which is the same as their input. Find the missing functions. Test your answers with various inputs in each case.

a $\longrightarrow$ [$\times 4$ -1] $\longrightarrow$ [?] $\longrightarrow$

b $\longrightarrow$ [-1 $\times 4$] $\longrightarrow$ [?] $\longrightarrow$

c $\longrightarrow$ [$\div 3$ $+2$] $\longrightarrow$ [?] $\longrightarrow$

C Mappings

This is a diagram to show the mapping:

$x \dashrightarrow x + 5$

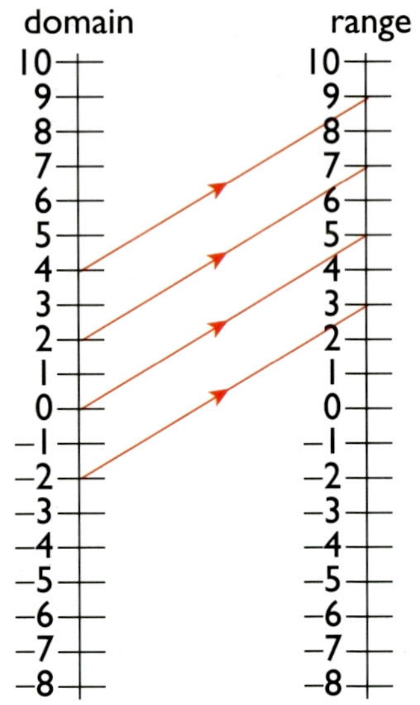

Numbers from one number line are the input (also called the **domain**) and they are mapped onto numbers on another number line (this is the output, also called the **range**).

C1 Draw diagrams to show the mappings below. Use tracing paper where you can to save drawing a lot of number lines. In some cases you may need to draw new number lines with different scales.

a $x \dashrightarrow x - 4$ **b** $x \dashrightarrow 2x + 1$ **c** $x \dashrightarrow x + 7$

d $x \dashrightarrow x^2$ **e** $x \dashrightarrow 10 - x$ **f** $x \dashrightarrow 2x - 5$

g $x \dashrightarrow 6 - 2x$

C2 What kinds of mappings produce diagrams with parallel lines like the example at the top of the page?

C3 Identify possible mappings in the following diagrams:

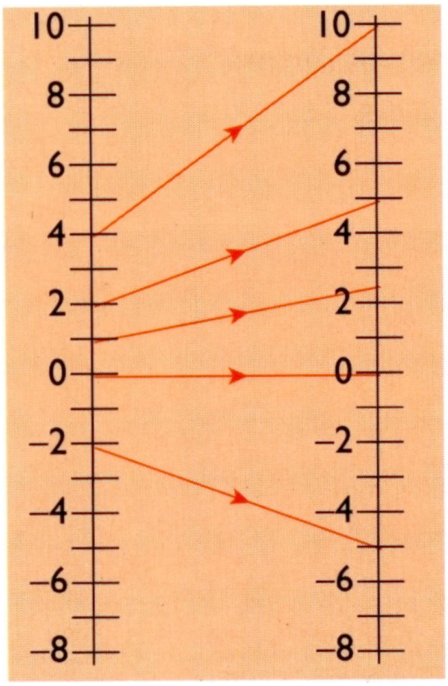

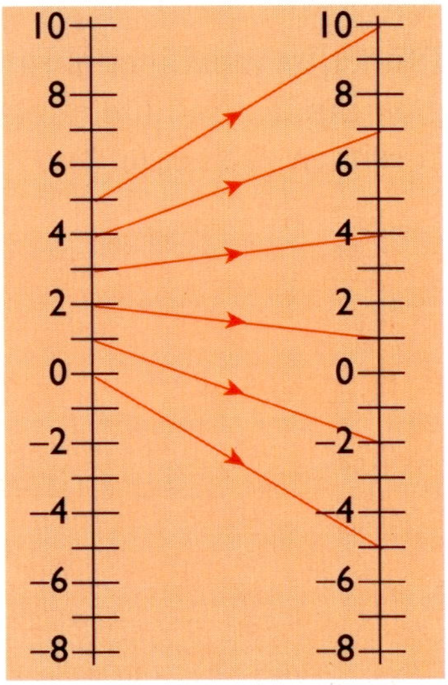

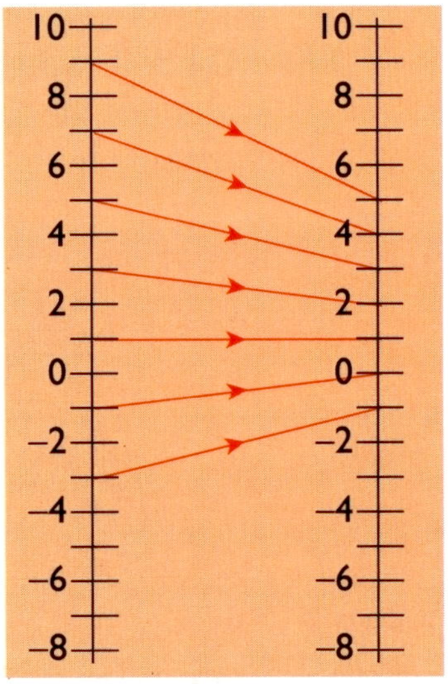

C4 What kinds of mappings produce diagrams where all the lines could be extended to cross at one point?

D Mappings and graphs

A graph is another form of mapping diagram. In diagrams like those on the previous page, the mapping $x \dashrightarrow x + 5$ can only be shown for a few numbers in the domain and range. Each pair of numbers needs a line to connect them. On a graph, each pair of numbers is represented by a point, marked on a grid with one axis (usually the x-axis) representing the domain and the other axis (usually the y-axis) the range.

Once the points are joined up by a line, then any other point on the line is also part of the mapping. It is important to consider whether this fits the situation that the mapping is describing.

Suppose a box of chocolates costs £3. A mapping diagram to show how much different numbers of boxes of chocolate cost could look like the one shown here.

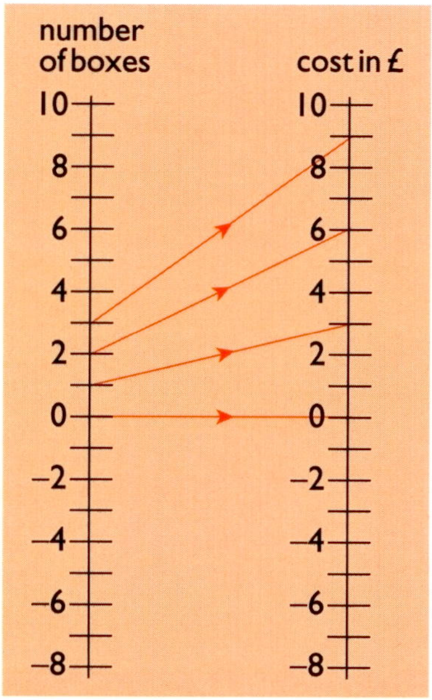

The mapping is $x \dashrightarrow 3x$
where x represents the number of boxes.

To make a graph you could plot the ordered pairs of numbers:

(0,0) (1,3) (2,6) (3,9) …

This is the graph of $y = 3x$, where y is the cost in £ and x is the number of boxes.

The points lie on a straight line. Should you join them up to make a solid line? If you do, you are suggesting that the mapping includes values of x such as 0.7, 1.25 or 1.995 and the corresponding values of $3x$. You obviously cannot buy 0.7 of a box of chocolates, so these numbers are not part of the mapping if it is representing a practical situation. The graph should therefore remain as a set of unconnected points.

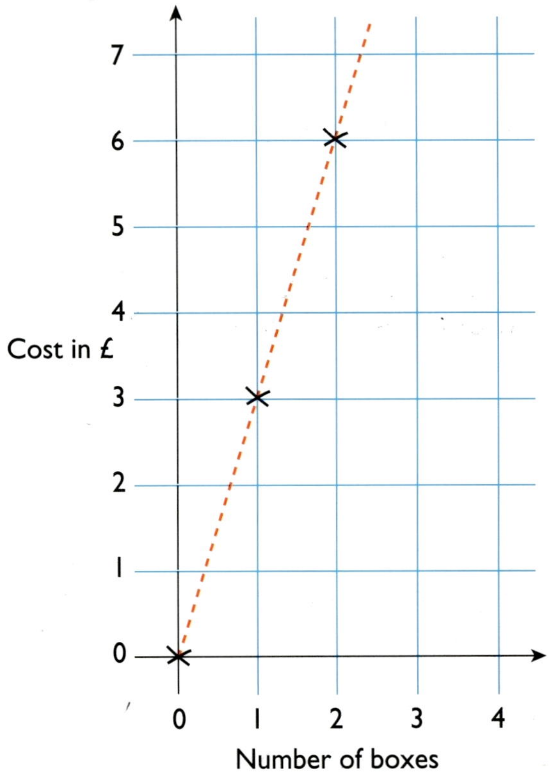

If the original situation had been about chocolate costing £3 per kilogram and you were buying it in small chips, then you could buy any quantity (more or less) and so it would be correct to join up the points on the graph.

D1 A shop sells packets of 2 m lengths of ribbon for 50p each. You can also buy the same type of ribbon for 20p per metre 'off the roll', in which case you can buy any length up to 50 m, which is the total length of the roll. Draw a graph to show the cost of ribbon up to 16 m bought

a in packets **b** off the roll.

Decide whether each graph should be a straight line or a set of spaced points.

D2 A filling station sells petrol from the pump at 50p per litre (minimum delivery 2 litres) and oil in 5-litre cans for £6 each. On the same axes draw graphs to show the cost of up to 20 litres of petrol or oil.

E Mapping ordered pairs

You can map pairs of numbers. Here is a coordinate grid with a quadrilateral marked on it. The vertices are at the points:

(2,2) (7,6) (6,1) (6,3)

The mapping $(x,y) \dashrightarrow (y,x)$ will swap the x- and the y-coordinates around to produce:

(2,2) (6,7) (1,6) (3,6).

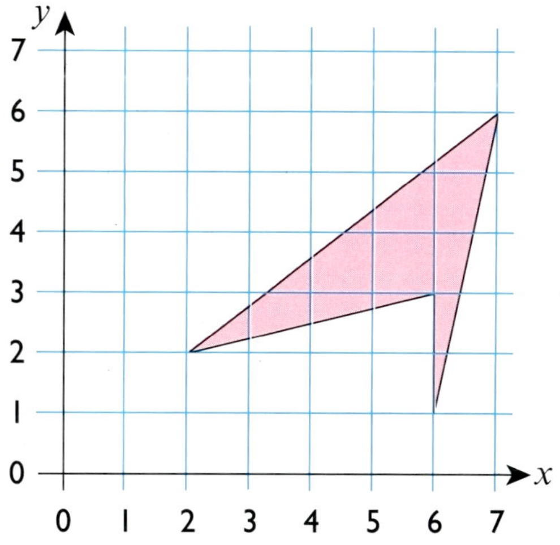

E1 Copy the diagram above and mark the new coordinates on the same diagram. Describe the mapping in geometrical terms (for example: translation, rotation or reflection).

E2 **a** Mark the shape shown here on a coordinate grid and show the following mappings on the same grid (work out before you start what scales you will need for the axes):

$(x,y) \longrightarrow (x + 5, y + 2)$

$(x,y) \longrightarrow (x - 4, y + 6)$

$(x,y) \longrightarrow (x, y - 5)$.

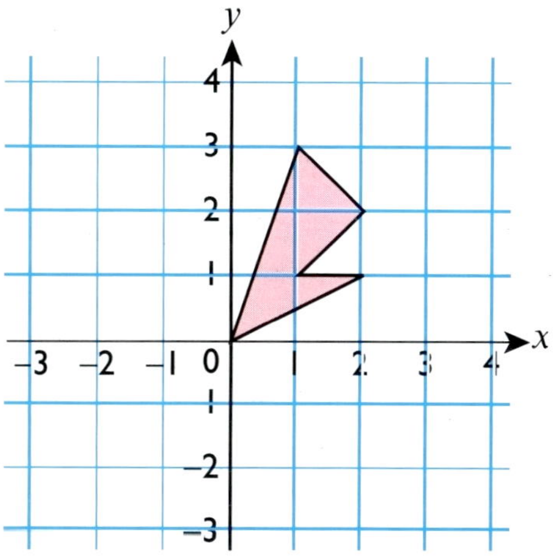

b Describe the mappings geometrically.

E3 On a grid make a triangle by joining the points (0,2), (2,3), (1,0). Try the following mappings and describe the effects geometrically:

a $(x,y) \dashrightarrow (2x,y)$

b $(x,y) \dashrightarrow (x,-3y)$

c $(x,y) \dashrightarrow (-x,-y)$.

E4 Mark the points (1,2), (1,4), (3,3), (2,1) on a coordinate grid. Find a geometrical interpretation for the following mappings:

a $(x,y) \dashrightarrow (6-x, 6-y)$

b $(x,y) \dashrightarrow (2-x, 2-y)$

c $(x,y) \dashrightarrow (-2-x, -2-y)$

d $(x,y) \dashrightarrow (6-x, -y)$

E5 These two coordinate grids show mappings of two objects. The two original objects are coloured in red.

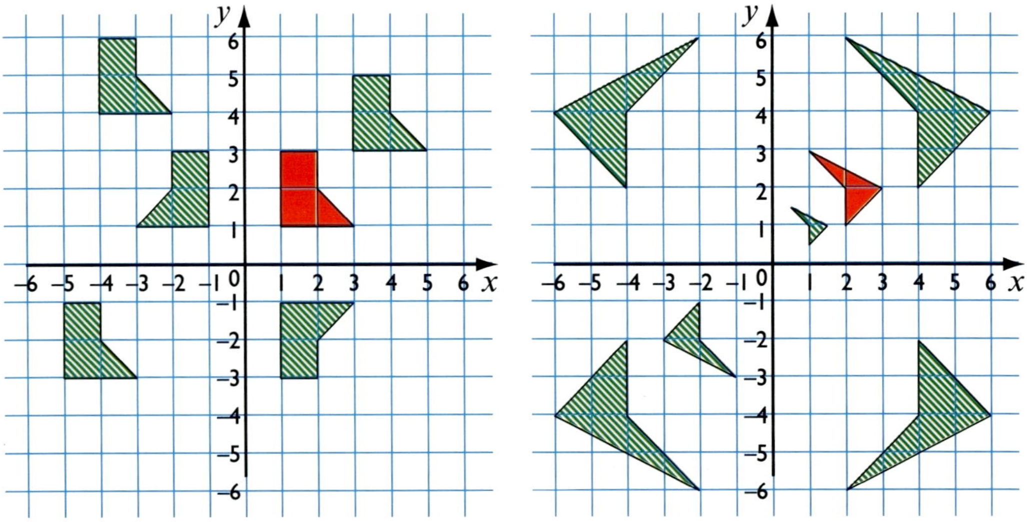

a Write down the mappings that produce each of the images of the original shapes.

b Find the reverse mappings that would transpose each image back to the original shape.

F A look ahead

Extension work

Not all graphs will be straight lines. Consider: $y = x^2$

F1 Use your calculator to help you complete this table:

$x =$	–5	–4.5	–4	–3.5	–3	–2.5	–2	... up to ...	4.5	5
$y = x^2 =$	25	20.25					4			25

Now choose an x-axis from –5 to +5 and a y-axis from 0 to 25 and plot the points as below.

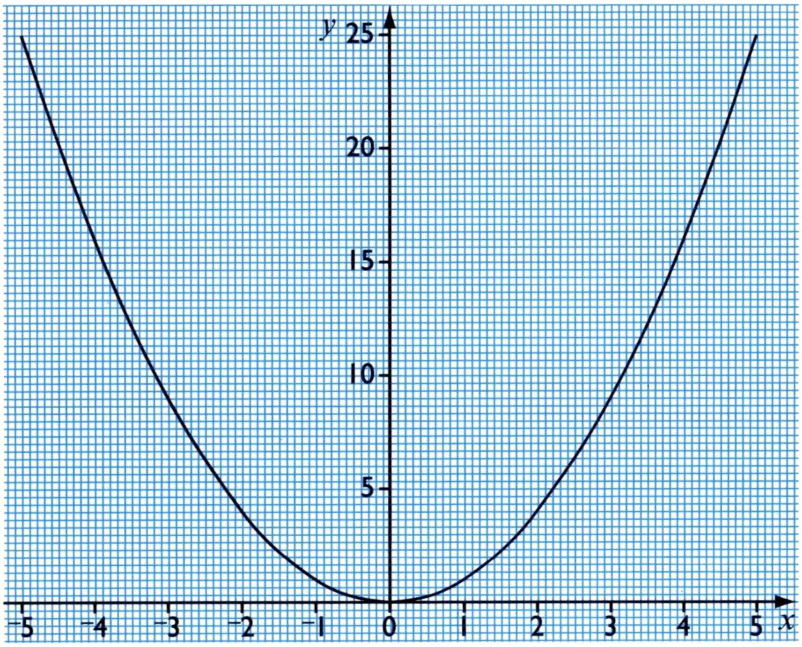

This shape is called a **parabola**.

F2 Using your graph find the squares of: 4.9 2.2 2.8 and compare your answers with those you get on your calculator.

F3 Now see if you can use the graph to find *square roots* of:

12 18 22 24.

Again compare with results from your calculator.

F4 Using the same method as in question **F1**, plot the graphs of:

a $y = x^2 + 1$ **b** $y = x^2 - 1$.

(Choose your own scales on the axes each time.)

UNIT 11 *Playing with numbers*

Numbers are useful – sometimes essential – in our everyday lives. They can also be fun. This unit contains activities and games – some you can try on your own while others will need one or more partners. You will probably not have enough time to attempt more than a selection of questions in this unit.

A Numbers in a square

A1 Copy this grid and put the numbers 1, 2, 3, ... ,16 in it in any order.

In *your* square, what is the total of all the numbers?

Write the sum of each row as a percentage of this total.

You will have something like this:

10	6	5	8
15	1	3	12
4	13	9	16
11	7	14	2

29 is 21.3% (correct to 1 decimal place)

31 is 22.8% (correct to 1 decimal place)

42 is 30.9% (correct to 1 decimal place)

34 is 25%

How can you check that you have got your percentages right? Do the check.

A2 When you fill in the square, what is the *lowest* total you could get for one row? What is the *highest*? Show how you get these totals.

Arrange the numbers so that the difference between the highest and lowest *row* totals is the same as the difference between the highest and lowest *column* totals.

A3 Arrange the same numbers, 1 to 16, in the square so that each row and column has the *same* total. What will the total be? Is there only one solution?

Can you also arrange for the *diagonals* to add up to the same total?

(When all the rows, columns and diagonals add up to the same number, you may remember from earlier work that it is called a **Magic Square**.)

A4 Repeat questions **A1**, **A2** and **A3** using a 5 by 5 grid with 25 squares using the numbers 1, 2, 3, ... , 25.

A5 If you have time, try using squares of other sizes.

B The game of numbers

For this game, you need two players. Make a 4 × 4 square with the numbers 1 to 16 written underneath like this:

1 2 3 4 5 6 7 8 9 10
11 12 13 14 15 16

You play alternately and, when it is your turn, you must enter one of the numbers in the square and cross it off. You must not enter a number which puts two consecutive numbers, such as 6 and 7, next to each other in any direction (vertically, horizontally or diagonally). You lose if you are unable to make a move.

Try some games and then answer the following questions.

B1 Is a draw possible? If so, how many ways can you find for getting a draw?

B2 If you play on a 3 by 3 square with the numbers 1 to 9, could you get a draw?

B3 Make a copy of the hexagonal board below or use Worksheet N3:5. Investigate a game on this board, using the numbers 1 to 15, with the same rule as before: you must not put any two consecutive numbers in adjacent spaces.

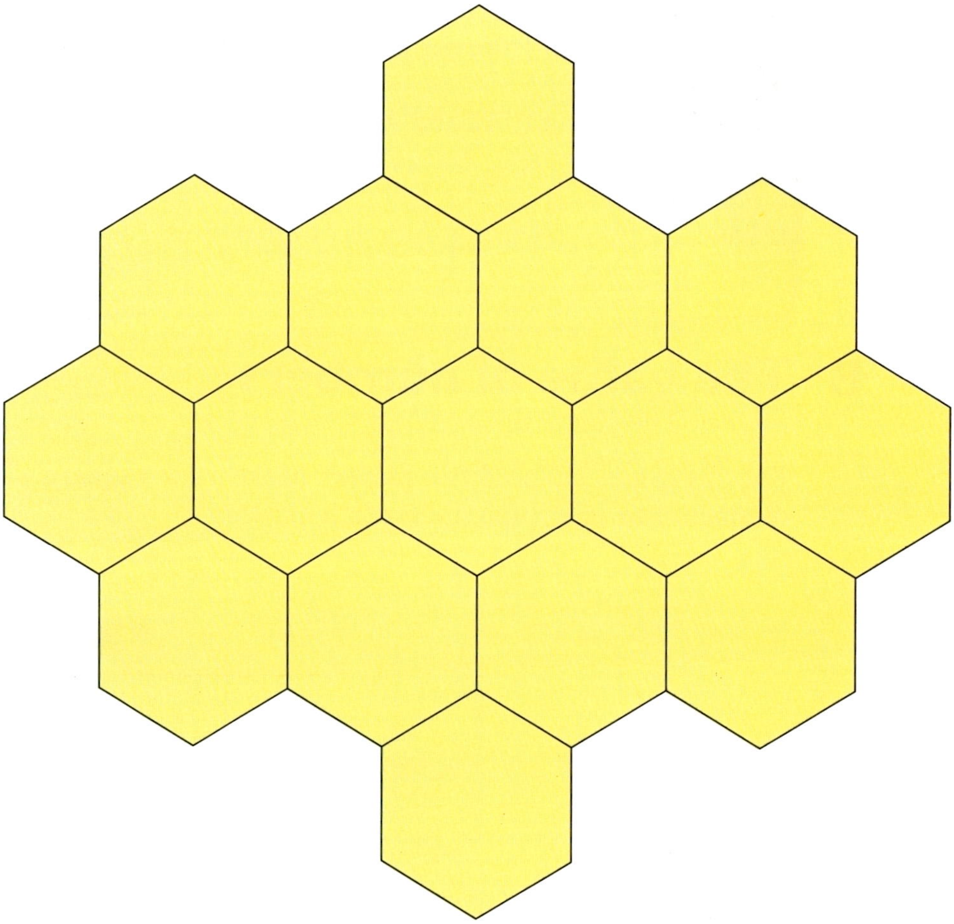

Extension work

B4 On a copy of the hexagonal board above, add four more hexagons to complete the outer ring of hexagons. Investigate a game using this board with the numbers 1 to 19. Do the four extra hexagons make the game more difficult or more interesting?

C Rows of numbers

1	2	3	4	5	6

Make a copy of the diagram. You will need a small counter or coin.

Place your counter on any number you choose and then move it to any other number adding the new number to the first. So, if you start at **3** and then move to **5**, you have a total of 8. Continue moving the counter and adding the new numbers. You have to try to finish with a total of 37.

C1 What is the quickest way to get to 37? What is the longest?

C2 Find if there is a way of getting to 37 without using **1** or **3**? Why could you not get 37 without using **1**, **3** or **5**?

C3 With a partner, play a game in which you each take turns to move the counter. The player who reaches 37 exactly, or forces the other player to go beyond 37, is the winner.

Work out how many ways you can win.

Make a new diagram with fractions. Again you will need a counter.

$\frac{1}{2}$	$\frac{1}{3}$	$\frac{1}{4}$	$\frac{1}{5}$	$\frac{1}{6}$

Follow the same rules as before. You might start on $\frac{1}{3}$ and then move to $\frac{1}{4}$ giving a total of $\frac{7}{12}$, and so on.

C4 Try to get a total of 5.
Find the quickest way and the longest way to get 5.

C5 Now try to get totals of 1, 2, 3, 4, $2\frac{2}{3}$, $4\frac{2}{5}$.

C6 Play a game with a partner in which you take turns to move the counter. The winner is the first to get exactly 5 or make the other player go beyond 5.

Now make another diagram with percentages:

5%	10%	20%	25%	50%

This time you start with 10 and you must try to get to a total of 100. Each time you put the counter on a square, you *add that percentage of your previous answer* to your total. For example:

You start with 10 (add 5%) 10.5 (add 10%) 11.55 (add 25%) 14.43 … . You will need to use a calculator and keep each answer to two decimal places.

C7 Can you get to *exactly* 100?

C8 What is the quickest way to get to a total greater than 100?

C9 Play a game with a partner in which you take it in turns to move the counter. The winner is the first to get 100 or make the other go past 100.

C10 Invent a game of your own which uses a combination of whole numbers, fractions and percentages.

D Number circles

Choose any whole number. Divide it by 3 if you can, if not add 5.Keep on doing this. What happens?

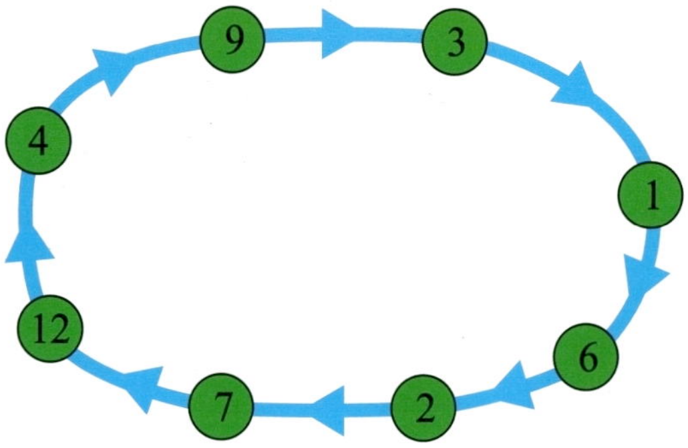

For example, if you start with 71 (which does not divide by 3 – how can you tell immediately?) you get:

71 76 81 27 9 3 1 6 2 7 12 4 *9 3 1 6 2 7 12 4 …*

You have got into a **number circle** *9 3 1 6 2 7 12 4* from which you will never escape.

D1 Try several different starting numbers and see what happens.

D2 Now use the rule 'divide by 7 if you can, if not add 5'.
See what happens.

D3 Make up different rules and investigate what happens.

E Some number patterns

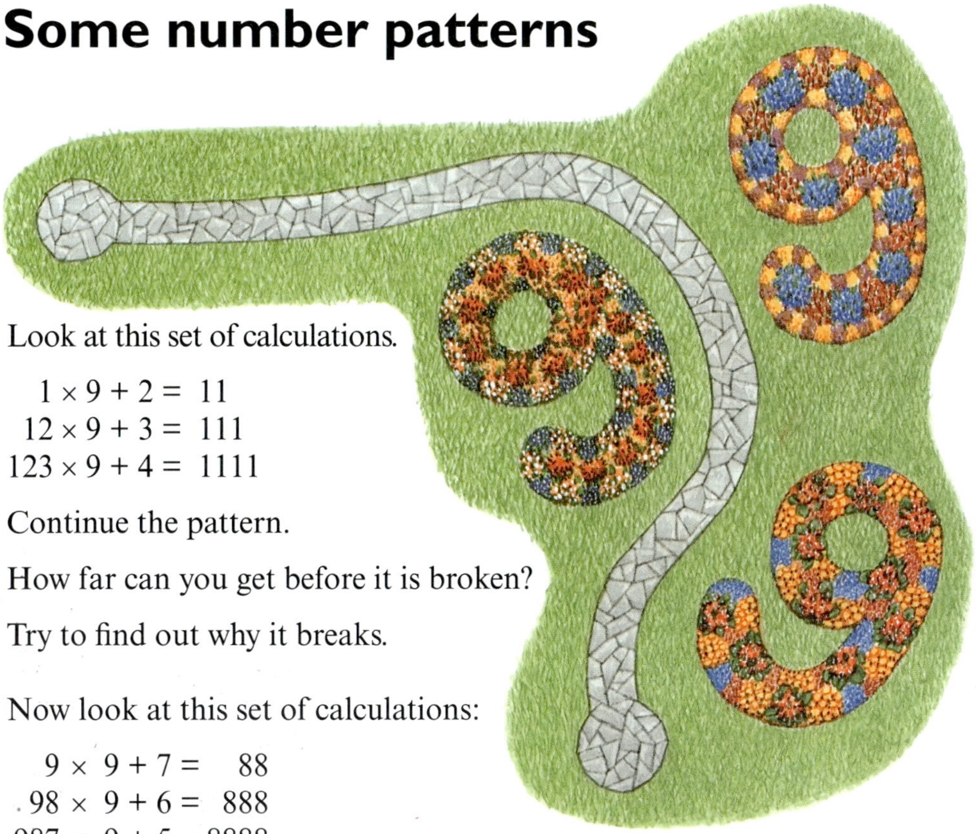

E1 Look at this set of calculations.

$1 \times 9 + 2 = 11$
$12 \times 9 + 3 = 111$
$123 \times 9 + 4 = 1111$

Continue the pattern.

How far can you get before it is broken?

Try to find out why it breaks.

E2 Now look at this set of calculations:

$9 \times 9 + 7 = 88$
$98 \times 9 + 6 = 888$
$987 \times 9 + 5 = 8888.$

Once again, continue the pattern and find out how far it goes before it is broken.

E3 Invent some similar patterns for yourself. Do not try to be too complicated. Multiplication by 9 often leads to interesting patterns.

E4 Find the square of the number 55446.
You may need to think carefully about this since the answer might be too large for your calculator. You may need to use a computer or pencil and paper.

When you get your answer, look at it carefully, as there is something special about it!
Find another square number with the same property.
(It might be a good idea to use a computer and the square root function in your search.)

E5 Write down your date of birth, for example 12.7.84.
Using just the numbers from that date, and using all of them each time, see how many numbers you can make starting from 1.
(Note: you may use a 2-digit number such as 12 either as *twelve* or as *one* and *two* separately.)
For example, using 12.7.84, we can make 1 in two different ways:

$1 = (12 \times 7)/84$ and $1 = (12 + 4) - (8 + 7)$

$2 = 8 + 4 - 7 - 2 - 1.$

E6 Can you arrange the digits 0, 1, 2, 3, 4, 5, 6, 7, 8, 9 to make 100?

Here is one complicated way: $50 + 49 + \frac{1}{2} + \frac{38}{76}$.

Check that this works and see if you can find another way for yourself.

E7 The Greeks thought that certain numbers were 'perfect'.
A perfect number is one which is equal to the sum of its factors, excluding the number itself.
(A factor is a number which will divide exactly into the number.)

For example, 8 has factors 1, 2, 4, but $1 + 2 + 4 \neq 8$ and so 8 is not a perfect number.

The number 6 has factors 1, 2, and 3, which is a perfect number because $1 + 2 + 3 = 6$.

What is the next perfect number after 6?

To find more perfect numbers, you could write a short computer program to help you in your search.

F Looking at numbers – mainly revision

F1 A small van can carry a maximum weight of 2000 lb.

The driver has to carry parcels weighing:

67 kg, 86 kg, 113 kg, 74 kg, 92 kg, 80 kg, 210 kg, 156 kg, 187 kg.

Can they all be carried together?
Show your working and use the fact that 1 kg is about 2.2 lb.

If he cannot take them all in the van, how many parcels (and which) should the driver leave behind? Assume that he wants to leave as few as possible.

F2 Without using a calculator, estimate:

a 145×19 **b** $367 \div 12$ **c** 498×15 **d** 119×49.

Now work out the exact answers using your calculator.
How far out were you as a percentage of each correct answer?

F3 A church wishes to raise £12 000 for repairs.

a In the first six months of the appeal, £2400 is collected. What percentage of the total is this?

b After 12 months, 55% of the total needed has been raised. How much is this?

c After 2 years, only £500 is needed. What percentage of the total has been collected?

F4

A 'megabike' costs £196 and is sold by shop A.
A 'maxibike' costs £210 and is sold by shop B.

Shop A has a sale giving 15% off everything.
Shop B has a sale giving 20% off everything.

Which bike is the cheaper and by how much?

F5 Use your calculator to do this calculation:

$$\frac{37.2 \times 123}{7 \times 3.8}$$

Write down the keys you press and the order in which you use them.

Now do the calculation using another method on the calculator. (Make sure that you get the same answer!)

What is the smallest number of keys you need to press to do it?

F6 A man buys a carpet for a rectangular room.

He finds that the room measures 15 paces one way and 13 paces the other. If the length of one of his paces is 78 cm, find the length, width and area of the carpet he needs.

If the carpet is sold in rolls 3 m wide, calculate what length of roll the man needs to buy and draw a diagram showing how he could lay the carpet.

F7 The population of a town is 55 500.

a If 12 000 people are over 45 years old, what percentage is this?

b If $12\frac{1}{2}\%$ of the population of the town are under 10 years old, how many is this?

c If the population increases by 15%, what is the new population?

d If the population increases from 55 500 to 60 000, what is the percentage increase?

F8 Here is some calculator practice. Correct your answers to 2 decimal places where necessary.

a 45.9×7.8

b $456 \div 4.34$

c $45.9 \div (56.7 - 34.3)$

d 35% of 456

e $\dfrac{34.5 - 67.8}{90.1 \times 2.34}$

f What percentage is 16 of 560?

g Decrease 600 by 12%

h Increase 500 by 54%

UNIT 12 Investigations and other activities

A Investigations

This Unit contains a variety of activities. You are not expected to work through all of them at the same time but, by the time you complete the whole book, you should have tried most of them. Keep a careful record of your answers.

A1 Number square square

1	2	3	4	5	6	7	8	9	10
11	12	13	14	15	16	17	18	19	20
21	22	23	24	25	26	27	28	29	30
31	32	33	34	35	36	37	38	39	40
41	42	43	44	45	46	47	48	49	50
51	52	53	54	55	56	57	58	59	60
61	62	63	64	65	66	67	68	69	70
71	72	73	74	75	76	77	78	79	80
81	82	83	84	85	86	87	88	89	90
91	92	93	94	95	96	97	98	99	100

Choose a square block of numbers, such as the one marked above, from a 10 by 10 number table. Multiply the opposite corner numbers together:

$23 \times 45 = 1035$ and $25 \times 43 = 1075$

Find the difference between the totals: $1075 - 1035 = 40$

Choose other squares from the table. Is the difference always the same when you multiply opposite corners together?

What happens if you change the size of the square or alter the size of the number table?

A2 Number search

You will need a calculator for this.

4 and 5 are **consecutive** numbers (which means they come next to each other in the counting sequence) and $4 \times 5 = 20$.

Can you find two consecutive numbers which will multiply to make 200?

If you cannot, then find the two consecutive numbers which will come closest to 200 when you multiply them. Write down your attempts as you make them. Continue until you are satisfied you have the best pair of consecutive numbers. Start a table like this:

Number	**20**	**200**	**2000**	**...**	**...**	**...**
best pair of consecutive numbers	4×5	??	??			

Continue by finding the two consecutive numbers that multiply to come closest to making 2000, then 20 000 and so on (up to 2 000 000 if you can).

What do you notice about the numbers in your table?

Can you make any predictions about what numbers you would get if you extended it?

Investigate the connections between $\sqrt{2}$ or $\sqrt{20}$ and the numbers in the table.

Can you explain them?

A3 Jugs

You have a supply of water and two jugs. The smaller jug holds exactly 3 litres when full and the other holds exactly 5 litres when full. There are no markings on the jugs and no other containers available.

How can you use the jugs to measure out whole numbers of litres up to 5 litres?

Invent a method of writing your answers in symbols.

Investigate other sizes of jugs.

A4 Rulers

This ruler is exactly 6 cm long. The only markings on it are at 1 cm and 4 cm from one of the ends.

How can it be used (with as few movements as possible) to draw lines which are any whole number of centimetres long up to 6 cm?

Devise a way of writing your answers using symbols instead of words.

Investigate the least number of marks needed on other lengths of ruler to be able to measure any whole number of centimetres.

A5 Chocolate eggs

A manufacturer of chocolate eggs produces two sizes: the Lunar which weighs 3 ounces and the Sonic which weighs 2 ounces.

He wants to sell a one pound box of eggs made up of a mixture of Lunar and Sonic chocolate eggs.

Investigate the possibilities.

(You should know that there are 16 ounces in 1 pound.)

When the manufacturer went metric, the Lunar became a 75 g egg and the Sonic became a 50 g egg.

Investigate the possibilities for a 500 g box of eggs.

The Market Research department reported that people like a box to contain at least 16 eggs but not weigh more than 1 kg.

In what ways could this be achieved?

A6 Square digit

Investigate the last digit of each of the square numbers (1, 4, 9 … and so on).

Can it be any number? Is there a pattern?

A7 Sixes and sevens

$6 \times 7 \div 2 = 21$
$66 \times 67 \div 2 = 2211$

Investigate this number pattern and find some others.

A8 Fivers

Investigate the answers to: 15^2 25^2 35^2 45^2 ……

From the number patterns in the answers, can you devise a short cut for squaring any number that ends in 5?

A9 Phoney

Do all the digits (0–9) occur with equal likelihood in telephone numbers? Investigate a page of the telephone book. What about car registration numbers and letters?

A10 Times 37

Investigate the 37 times table for number patterns.

A11 Stamps

At the local Post Office, they ran out of the usual first and second class stamps. They only had 5p and 8p stamps left.

The counter clerk said that it didn't really matter, as you could make up the price of a first or second class stamp using only 5p and 8p stamps or a mixture of the two kinds.

Was that true when a first class stamp cost 24p and a second class stamp cost 18p? If so, show how it could be done.

Was it true when the first class stamp went up to 25p and the second class stamp went up to 19p? If so, show how it could be done.

The clerk went further and claimed that you could make up any value of postage from 28p upwards and quite a few of the amounts less than 28p, just using 5p and 8p stamps.

Investigate whether that is true.

Investigate what the situation would be if you only had 4p and 7p stamps.

A know-all mathematician came into the post office and mumbled:

'Mmm… quite interesting really… if they only have stamps for xp and yp left then you can make up all the values from $(x-1)(y-1)$p upwards… except when…'

Her mumbling was then indistinct. Investigate her statement.

A12 $4u + t$

Write down a two-digit number such as 35. 'u' is the units number and 't' is the tens number. In this case $u = 5$ and $t = 3$.

Work out $4u + t$: $4 \times 5 + 3 = 23$.

Repeat the process with the 23: $u = 3$ and $t = 2$.

Work out $4u$ + t: $4 \times 3 + 2 = 14$.

So far you have 35 ⟶ 23 ⟶ 14 ⟶ ?

Continue the process and see what happens.

Try different numbers and investigate whether they all behave in the same way.

Can you find a number such that $4u + t$ equals the number? Can you find a way of predicting the number instead of simply trying different ones?

What about $5u + t$ sequences, or $2u + t$?

A13 Diagonals

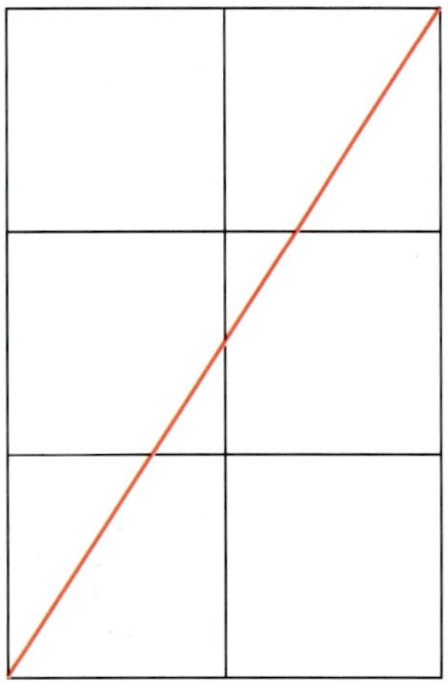

If you draw a diagonal through a block of squares, it passes through some of the squares.

What is the connection between the number of squares it passes through and the length and breadth of the block of squares?

A14 Biggest box

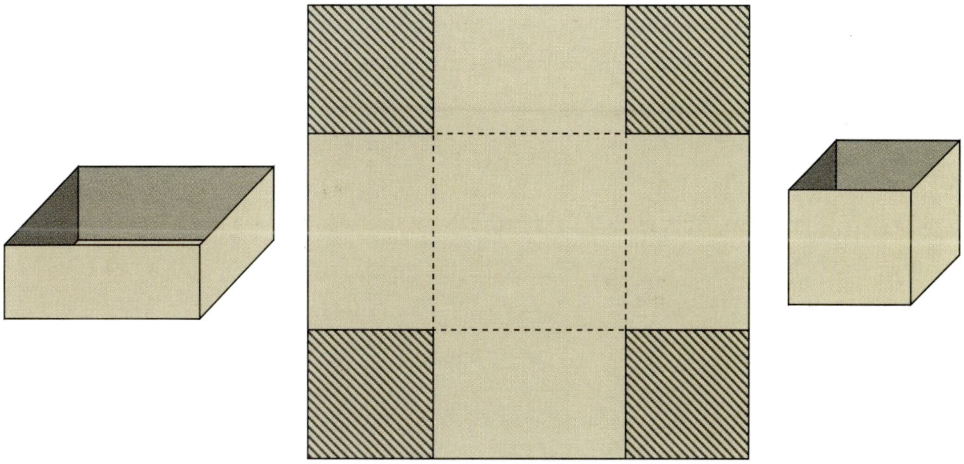

You can make an open box from a square of paper by cutting four equal corner pieces out of the square and folding up the edges. These edges can then be taped to make the box.

The size of the box will depend on how big the corner pieces that are removed from the original square are.

Suppose you started with a square of paper 20 cm by 20 cm. Investigate how the volume of the box varies with the size of the corner pieces cut out.

What measurements do you need to make the box with the biggest volume?

A15 Just the same

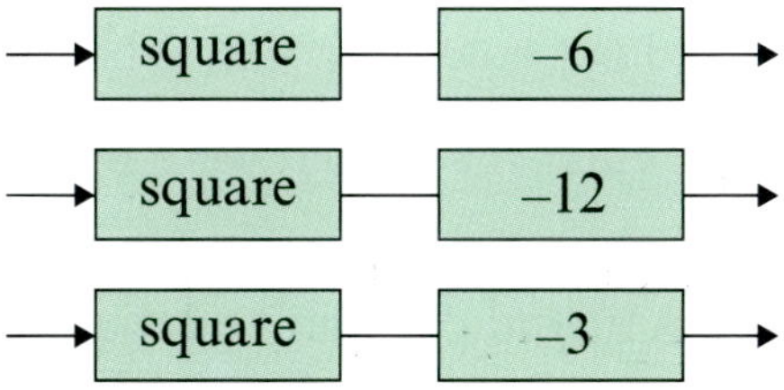

Each of these pairs of function machines can be made to supply an output which is the same as the input – if you find the right input numbers.

Use a trial and improvement method in each case to find suitable inputs.

B Playing with spreadsheets

B1 **a** On a spreadsheet enter the following sequence:

spreadsheet

	A
1	3
2	7
3	11
4	15
5	19
6	23
7	27
8	...
9	...
10	...

A
3
= A1 + 4
= A2 + 4
= A3 + 4
= A4 + 4

Put a number into cell A1 and then, using a formula which connects A2 with A1, and so on, generate the rest of the sequence. The number of the term is shown by the number of the spreadsheet. By altering the data in A1 can you get 50 in cell A10? What number would you then get in A20 and A30?

b By only altering the data in A1, try to get each of the following numbers in cell A10: 33, 981, 3, 7.2, 9.13, –2, –1.1.

Write down how you worked out how to do this. What clues did you use?

Did you develop any short cuts? If so, what were they?

c What changes do you have to make to the number in A1 if the spreadsheet has 50 in cell A10 and you want to change it so that 50 appears in cell A12?

d What changes do you have to make to the number in A1 if the spreadsheet has 50 in cell A10 and you want to change it so that 50 appears in cell A15?

e What about 50 appearing in cell A20, then A23, then A31, then any A cell? Describe as carefully as possible any rules which you discover.

B2 Try the activities in question **B1** using these sequences:

a 3 8 13 18 23 28 33 38 43

b 1 10 19 28 37 46 55 64 73

c 6 6.2 6.4 6.6 6.8 7 7.2 7.4 7.6

B3 **a** Put the sequences represented by these formulae on to a spreadsheet and answer the questions below:

	A	B	C	D	E
1					
2	=A1–3	=B1–2	=C1–1	=D1+1	=E1+2
3	=A2–3	=B2–2	=C2–1	=D2+1	=E2+2
4					

b What numbers do you put in cells A1, B1, C1, D1, and E1 so that the numbers in the cells in row 2 are:

i 5 **ii** 10 **iii** 50 **iv** –6?

c What numbers do you put in cells A1, B1, C1, D1 and E1 so that the numbers in the cells in row 11 are:

i 31 **ii** 53 **iii** –27 **iv** 53.1?

d How can you work the answers out without using the spreadsheet?

B4 On a spreadsheet, design sequences with:

a 7 in A1 and 23 as a term

b –5 in A1 and –11 as a term

c –3 in A1 and 5 as a term

d 0.5 in A1 and 6 as a term.

Number and Algebra 3
National Curriculum level description coverage

NC level and description	Covered in pupil's book unit:
• order and approximate decimals when solving numerical problems and equations such as $x^2 = 20$, using trial-and-improvement methods	1
• to be aware of which number to consider as 100 per cent, or a whole, in problems involving comparisons, and use this to evaluate one number as a fraction or percentage of another	2, 3, 6, 7, 11
• understand and use the equivalences between fractions, decimals and percentages, and calculate using ratios in appropriate situations	2, 3, 6, 7, 11
• when exploring number patterns, to find and describe in words the rule for the next term or *n*th term of a sequence where the rule is linear	4, 5, 12
• formulate and solve linear equations with whole number coefficients	9
• represent mappings expressed algebraically, interpreting general features and using graphical representation in four quadrants where appropriate	8, 9, 10
• using and applying the above skills	All units especially 11 and 12